Teresa J. C. Welch, M.S.

Assistant Professor, Department of Chemistry
Lindenwood College
St. Charles, Missouri

E. James Potchen, M.D.

Professor of Radiology
Director of Nuclear Medicine
Mallinckrodt Institute of Radiology
Washington University School of Medicine
St. Louis, Missouri

Michael J. Welch, Ph.D.

Associate Professor of Radiation Chemistry in Radiology
Mallinckrodt Institute of Radiology
Washington University School of Medicine
St. Louis, Missouri

FUNDAMENTALS OF THE TRACER METHOD

1972 — W. B. Saunders Company

PHILADELPHIA • LONDON • TORONTO

W. B. Saunders Company: West Washington Square
Philadelphia, Pa.   19105

12 Dyott Street
London, WC1A   1DB

833 Oxford Street
Toronto 18, Ontario

Fundamentals of the Tracer Method   ISBN   0-7216-9180-3

Print No.:   9   8   7   6   5   4   3   2   1

*THIS BOOK IS DEDICATED TO OUR
PARENTS AND TEACHERS*

PREFACE

In modern medicine and biology there is an ever-increasing shift of emphasis from static observations to kinetic considerations. This has in part been brought on by the use of tracers — dyes, nonradioactive and radioactive isotopes — to study normal physiologic and abnormal pathologic events. The change of emphasis in research and practical work has required contemporary physicians, biologists, biochemists, and physiologists to obtain a more thorough understanding of the tracer method.

The purpose of this book is to serve as a basic introductory text for graduate and undergraduate students, as well as a review for investigators and researchers who are already knee-deep in tracer techniques. It could be used in a wide variety of circumstances: as a text for a meaningful segment of a biomathematics course, as a basis for beginning and designing a research project, as a basic text for radiologists and nuclear medicine specialists, and as a review book.

Fundamentals of the Tracer Method ranges from an introductory definition of a tracer to the airing of some current theory that is likely to involve subsequent generations of investigators. It carefully defines tracers and tracer techniques, demonstrates the application of appropriate mathematical and statistical tools, covers the general design of an experiment and data presentation, and follows through with a thorough discussion of the dynamics of tracer studies and their interpretation.

It is generally accepted that many physicians trained in the biologic sciences have had insufficient instruction in mathematics to grasp some of the concepts presented in modern journals and books. This non-recognition of the potential of some very fundamental but powerful mathematical tools has prevented the wider application of the principles of the tracer method to practical research and clinical problems. With this situation in mind, we have attempted to direct the book toward students in medicine and biology who are not thoroughly grounded in mathematical fundamentals. In fact, familiarity with mathematical theory is not required, since the application of the concepts rather than their derivation is emphasized.

Research with tracer techniques is fraught with many statistical problems and requires consideration of errors in experimental design. Thus, we have attempted to develop a systematic approach to some experimental problems that are potentially solvable by the tracer method. It is hoped that this applied approach to the tracer method, with frequent examples, will prove helpful to students, investigators, and clinicians alike. We seek the indulgence of our more sophisticated and mathematically-oriented colleagues, who may be offended by applied mathematics without formal structuring.

TERESA J. C. WELCH
E. JAMES POTCHEN
MICHAEL J. WELCH

ACKNOWLEDGMENTS

The authors acknowledge the generous assistance of our colleagues and associates, especially Dr. Mustafa Adatepe, Dr. Alexander Gottschalk, and Loretta Morgan.

CONTENTS

Chapter 1

TRACERS

Complex disciplines, such as the Tracer Method, require systematic, structured thought; such thinking is a participant and not a spectator sport. The participant must actively evaluate, manipulate, and apply the appropriate techniques if he is to become facile in such a thought game. Therefore, it has been the authors' objective in this book to outline the discipline and to supply the rules — definitions, restrictions, tools — encouraging the reader to structure his application in relation to his needs and past experience.

The authors realize that it is an ambitious endeavor to attempt a comprehensive discussion beginning with the most basic definitions and ending with sophisticated experimental techniques and current theoretical thought. However, if the book is to be useful, it must carry the reader to the stage of tracer work now underway, in which he is likely to be involved. Recognizing that most readers will be allied with some medical specialty or biological subject, the authors have made free use of biomedical examples to demonstrate techniques and to alert the reader to problems he likely encountered at some time in his past experience.

1.1 WHAT IS A TRACER?

The "tracer method" has gained prominence through broad application to many disciplines, from medicine to biology to engineering. In its simplest definition *a tracer follows or outlines something else, the tracee.* The tracer, a labeled species, must act the same as the nonlabeled species in the system of interest, and its behavior must be representative of the whole population of the species. Although the tracer may be made of the same "stuff" as the tracee, this is not essential. The tracer must only act like the tracee. When the tracer method is applied to biological or chemical processes, it is necessary to impose another, perhaps evident, restriction; namely, the tracer must not interfere with the process being studied.

The tracer must somehow be labeled, must engage in the process of interest, and must be representative of the tracee. The label or tag we pin on a tracer may be radioactivity or a special dye molecule or any other detectable material.

Tracers which are commonly used in biomedical studies may be divided into three categories: dyes, nonradioactive (stable) isotopes, and radioactive isotopes.

1.2 DYES

Dyes are chemical tracers; that is, it is the special chemical properties of dye molecules that allow us to trace them. The dye is, therefore, detected by chemical means. Dyes such as cardiogreen and bromsulphalein (B.S.P.) are used in tracer studies when one wishes to determine the size of a population, the amount of substance, or the number of objects from only a small sample that is accessible to observation. These determinations are termed tracer dilution methods and will be discussed in greater detail in Chapter 6.

Spectrophotometry. Spectrophotometry is the conventional chemical method used to determine dye concentrations. This analysis depends upon the measurement of electromagnetic radiation, at a particular wavelength, absorbed by a sample. When light (electromagnetic radiation) is passed through a translucent substance, the amount of radiation absorbed or transmitted will be dependent upon the amount of substance present. The experimental quantity measured in a spectrophotometer is usually either transmittance T or absorbance A. Transmittance is a ratio between the amount of light striking the sample and the amount passing through the sample ($T = I/I_0$, where I_0 is the initial intensity of the light beam and I the transmitted intensity). Absorbance A is the negative logarithm of transmittance:

$$(1.1) \qquad A = -\log T = abc$$

where a is simply a constant that depends upon the experimental conditions as well as the nature of the absorbing material; b is the path length, that is, the length of material through which the light beam must traverse; and c is the concentration of the sample. If b is expressed in centimeters and c in grams per liter, then a is called

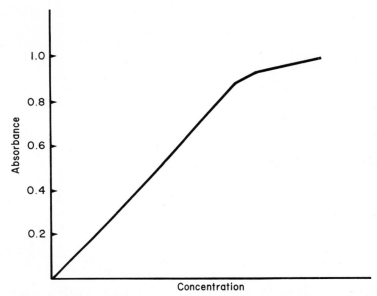

Figure 1–1. Absorptivity is given by the slope of the plot of absorbance against concentration. It should be noted that at high concentration the plot deviates from linearity (i.e., deviates from straight line).

the absorptivity. If c is given in moles per liter, then the constant a is the molar absorptivity (or extinction coefficient) and is given the symbol ϵ. Since A is then directly proportional to b and c, the use of the constant (extinction coefficient) avoids the more complicated exponential function.

Equation 1.1 suggests that the determination of absorbance allows the calculation of the sample concentration, provided a and b are known. Since most samples are measured in standard cells of known dimensions, b is known. The absorptivity a, however, should be determined for each instrument, because different spectrophotometers have different characteristics. The a may be determined by measuring the absorbances of solutions of known concentration and plotting A versus c, or bc, the slope* of which is a (Fig. 1–1).

Determining concentration by spectrophotometry is usually restricted to the visible and ultraviolet regions of the electromagnetic spectrum; that is, the concentrations of ions or molecules that absorb strongly in the visible or ultraviolet regions may usually be determined by this method. Spectrophotometric determinations may also be made on strongly absorbing derivatives of the substance of interest. For example, suppose some material X does not absorb well, but X

---

*Slope of a line is measured as the rate of change of the ordinate with respect to the abscissa.

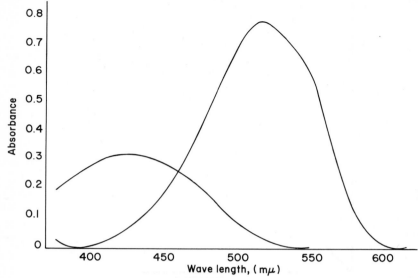

Figure 1-2. Absorption spectrum of a typical dye in which peak heights may be used for analytical determination. Peaks are nearly symmetric; i.e., there are no "shoulders" to the curves.

may be converted quantitatively to XY, which is highly colored. The concentration of X may be obtained by determining the XY absorbance. Figure 1-2 shows a typical absorption peak which may be used for analytical determinations.

A note of caution must be sounded about the ease of spectrophotometric methods; many systems deviate from the simple linear* relationship given in Equation (1.1). Very concentrated solutions, for example, usually deviate from linearity. It is, therefore, reemphasized that calibration of the instrument in the concentration range of interest is essential.

Example. Suppose the absorbance of a dye solution, with an absorptivity of 110, is found to be 0.541 in a cell of length 1 cm. The concentration of the solution can be found simply by substitution in Equation 1.1:

$$0.541 = (110) \, (1) \, c$$

$$c = \frac{0.541}{(110)} = 0.00499 \text{ g/l}$$

---

* Linear can be defined three ways: (1) in a straight line. (2) Along a curve. (3) Having only one direction.

1.3 NONRADIOACTIVE TRACERS

Some nonradioactive isotopes of elements with special physical properties may be used as tracers. In these cases, it is the special physical property of the isotope that is the label or tag. Two non-radioactive tracers that are commonly used in biomedical studies are deuterium, an isotope of hydrogen with a mass of 2, and carbon 13, an isotope of carbon with a mass of 13. Deuterium may be used as a tracer in the form of D_2O, the deuterium analog of water, H_2O. Since no radioactivity is involved, these isotopes must be detected by chemical or physical means. The two methods commonly used are mass spectrometry and nuclear magnetic resonance (NMR).

Mass Spectrometry. Mass spectrometry is a method by which the molecules of a sample are converted to ions and these ions sorted according to their mass (more precisely, to their mass-to-charge ratio). Nearly every substance that is a gas or can be converted to a gas may be analyzed by means of a mass spectrograph. The molecules are ionized (usually by bombarding with a stream of electrons) and accelerated to form a beam of ions. This beam is subjected to a magnetic field in which the initial beam is sorted into separate beams of different mass, based on the principle that the greater the mass of an ion, the less deflection of that ion in a magnetic field. A suitable detector locates and measures the separated ion beams. Deuterium compounds may be detected by this method. For example, the isotopic composition of hydrogen gas may be determined by measurement of mass 2, 3, and 4 intensities, which correspond to ionized H_2, HD, and D_2 (the mass of hydrogen is 1; of deuterium, 2). Figure 1–3 shows a schematic diagram of a mass spectrograph.

Nuclear Magnetic Resonance. Nuclear magnetic resonance may be used to detect certain nuclei (normal hydrogen of mass 1, carbon

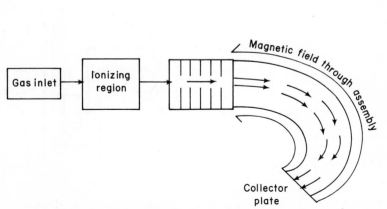

Figure 1–3. Schematic diagram of a mass spectrograph.

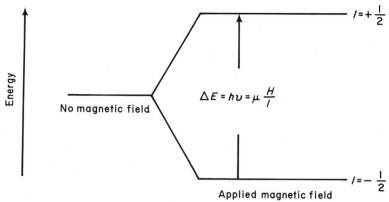

Figure 1-4. Splitting of nuclear spin state under the influence of an applied magnetic field. ΔE is the energy associated with the transition between the lower energy state $I = \frac{1}{2} + 0$ and higher energy state $I = +\frac{1}{2}$.

13, nitrogen 15, and fluorine 18) that act as though they were charged spinning bodies. When subjected to a magnetic field, these nuclei have an orientation in which they are effectively lined up with the field or against the field. The nuclear magnetic resonance (NMR) method is simply another kind of spectroscopy; the incident radiation is in the radio-frequency region, and some nuclei are induced to change from the more favorable, lower energy state in which they are aligned with the field to the less favorable, higher energy state opposed to the field. Such a transition is shown in Figure 1-4.

If a sample is subjected to a constant magnetic field H_0 and irradiated with radio-frequency energy, not all the nuclei of the same element will have the same resonance frequency, because their magnetic fields are also determined by electrons around the nucleus. Nuclei in different chemical environments will have different elec-

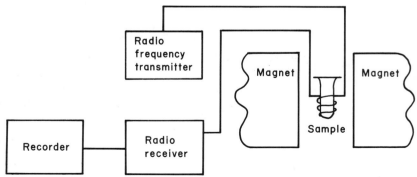

Figure 1-5. Schematic diagram of a nuclear magnetic resonance spectrometer.

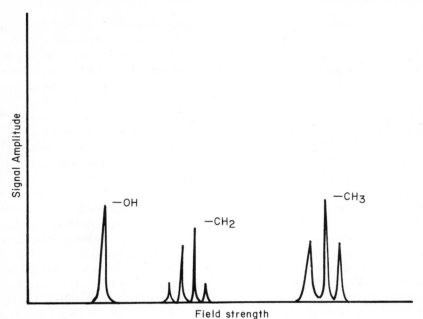

Figure 1-6. NMR spectrum of ethanol under high resolution.

tron distributions and therefore, experience different magnetic fields. The effective magnetic field that a nucleus "sees" is not H_0 but H_0 $(1 - \sigma)$, where σ is a parameter characteristic of a particular chemical environment for the nuclei. This change in resonance frequency when scaled to some standard frequency is called a *chemical shift*. The chemical shift serves as an analytical probe for subtle changes in the effective environment of a nucleus.

Figures 1-5 and 1-6 show a schematic of an NMR spectrometer and a typical NMR spectrum, respectively. The area under each curve is proportional to the amount of substance present.

1.4   RADIOACTIVE TRACERS

Reactions studied in chemistry and biochemistry involve transformations of the electrons (usually valence electrons) outside the nucleus. The nucleus itself retains its identity. In contradistinction, those processes that are collectively known as "radioactive" involve transformations of the nucleus. Most chemical transformations involve energies of about 10^4 to 10^5 cal/mole of reactant, whereas energies of nuclear processes are of the order of 10^{10} cal/mole of reactant.

Radioactivity occurs in nuclei that are unstable; that is, they are in excited energy states. Decay allows the nuclei to reach a lower energy (more stable) state. The radiations emitted from radioactive substances consist of three principal types: alpha (α), beta (β), and gamma (γ). The relative energies, and hence penetrating powers, of these radiations are $\gamma > \beta > \alpha$. When these high-energy radiations are absorbed by tissue or a detector, they cause ionization of some of the atoms making up the absorber, hence the term *ionizing radiations*. An alpha particle is simply a high-velocity helium nucleus (i.e., two protons, two neutrons, and no electrons). A beta process may result in the emission of a high-velocity electron called a beta-minus or the emission of a high-velocity positron (same as an electron but with positive charge) called a beta-plus. A gamma transition involves simply the emission of a quantum of electromagnetic energy.

Because of their detectability, radioactive isotopes may be suitable as tracers. Usually those isotopes that emit beta and gamma radiation are used, because alpha radiation is not sufficiently penetrating to be readily detected. Some examples of radioisotopes currently used as tracers are (1) beta emitters — carbon 14, tritium, phosphorus 32; (2) gamma emitters — iodine 131, technetium 99m, chromium 51, mercury 197, indium 113m, and so forth. The use of radioactive tracers involves a compromise between energy (penetrating power of the particular decay process), half-life* and the tracer potential of the isotope.

1.4.1   Measurement of Radioactive Tracers

Radioactive tracers are measured by various detectors; the three types to be discussed here are gas-filled counters, scintillator-type counters, and solid-state counters. Of the three types the gas-filled counter is the least expensive but is of little use to detect gamma radiation; the solid-state detector, usually used in high-energy physics, is the most sensitive type.

1.4.1.1   Gas-Filled Detectors

A simple gas-filled detector is shown schematically in Figure 1–7; the inner wire and the case act as two electrodes. This simple counter can be used as either an ion chamber, a proportional counter, or a Geiger-Müller counter, depending upon the voltage applied. At low voltage a first plateau is reached (Fig. 1–8) when all the ions caused by the nuclear event are collected by the center wire. In this region only the primary ionization is collected, and any required amplification must be done in the allied electronics since there is no amplification in the counter itself.

---

*Half-life is specifically defined as the time during which half of the particular radioisotope decays.

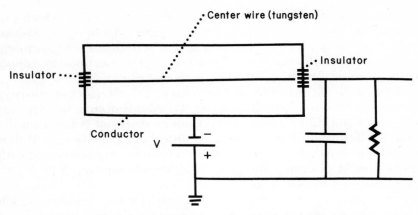

Figure 1-7. Schematic diagram of a simple gas-filled detector.

As the voltage is increased, the ionized products in the absorber (gas) acquire energy, and additional ionization occurs. Throughout this "proportional region" each initial electron produces N secondary electrons. N is the multiplication factor of the counter. This region is termed the proportional region because the size of pulse (i.e., voltage deflection) caused by each ionizing event is proportional to the number of ions formed initially.

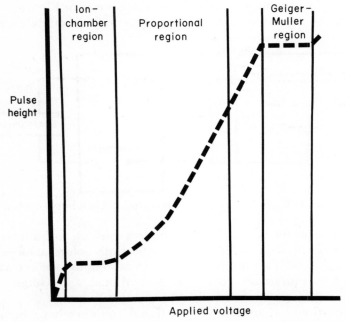

Figure 1-8. Variation of pulse height with applied voltage, indicating operating regions for various types of detectors.

A pulse-height discriminator is an electronic device which can separate the incoming pulses by their energies. When a proportional counter is coupled to a pulse-height discriminator, high-energy electrons can be distinguished from low-energy electrons by adjusting the discriminator to reject low-energy pulses.

Eventually, as the voltage is increased, a single electron initiates ionization in the whole of the counter. When this occurs, the detector produces a pulse of constant height which does not depend on the initial ionization. In this region the counter is called a Geiger-Müller or simply a Geiger counter. This type of counter cannot be used with a pulse-height discriminator because all events registered by it produce a pulse of the same size.

Although any gas counter can be used for beta counting, each counter has advantages in certain situations which will be discussed separately.

Ionization Chambers. When electrons and positive ions are formed by the action of the nuclear particle on the gas, their behavior depends on the nature of the gas and the electric field present. In an *ion chamber* the field is such that the electrons move to the wire. The ion chamber, like all gas chambers, has very low efficiency for gamma radiation owing to the fact that very few of the gamma rays are stopped in the gas. The current produced by the ions is measured by using an electrometer capable of detecting typical current of approximately 10^{-10} A. A schematic of an ion chamber is shown in Figure 1–9. The main use of ion chambers is to measure high amounts of activity.

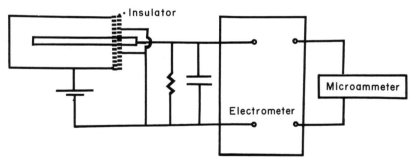

Figure 1–9. Schematic diagram of an ion-chamber detector.

Since the counter is simply measuring a current, not a count rate, one does not have the "dead time" (the time before another count can be observed after an event occurs) problem associated with the counting of individual pulses in proportional and Geiger-Müller counters.

Proportional Counters. Proportional counters have been used in various shapes and sizes for many different applications. Figure 1–10 shows two simple counters: (a) is a loop proportional detector

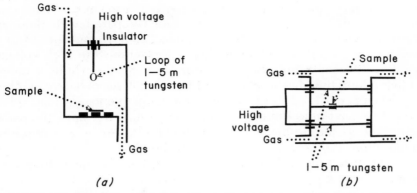

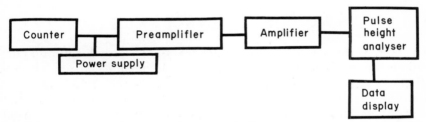

Figure 1–10. Schematic diagram of proportional counters. (*a*) Loop proportional detector for counting planchet samples; (*b*) 4π counter where sample is placed between two detectors.

used to count planchet-type samples; (b) is a 4π counter where the sample is placed between two separate counters. In this way radiation emitted in every direction (that is, a 4π solid angle) is counted. Many types of gases and gas mixtures can be used in proportional counters; typical gas mixtures are helium-methane, helium-propane, and argon-methane.

Figure 1–11. Schematic diagram of the components used in pulse-height discrimination.

Proportional counters have a low dead time; over one million counts a minute can be accumulated without significant count loss. As the pulse height is proportional to the energy of the actual event, pulse-height discrimination can be applied to the output from proportional counters. A typical setup for pulse-height discrimination is shown in Fig. 1–11. Such discriminator circuits are frequently used to count double-labeled samples from chromatographic columns, especially carbon 14 and tritium-labeled compounds. Such compounds are often counted in the gas phase since they can easily be converted into carbon dioxide and water, with the activity being passed directly through the counter in the stream of gas.

Geiger-Müller Counters. Geiger-Müller counters were very popular a few years ago; they have the advantage of being inexpensive and very rugged. They have, however, a major disadvantage; the maximum counting rate is low when compared with those of other detectors. This is due to a long dead time. In a Geiger-Müller counter the absorbed radiation causes ionization; the high voltage amplifies the ionization, causing the pulse to fill the whole chamber so that a long time has to pass (dead time = 250 μsec) before another pulse can occur (Fig. 1–12). Even if radiation events are spaced beyond the dead-time limit, the initial pulse may not have the same amplitude as the original one, and an even longer time is required before there can be a pulse of the initial height.

This dead time is shortened by the addition of a "quenching gas" to the counting gas (commonly 0.1 percent chlorine). The inert-gas ions collide with the quenching-gas molecule on the way to the cathode. The charge on the inert gas will frequently be transferred to the quenching gas, whereas the reverse is not possible. Thus, ions reaching the cathode are quenching-gas molecular ions that become neutralized. If only inert-gas ions reached the cathode, they would initiate secondary emission, and the duration of ionization, and hence the dead time, would be longer. If the dead time for a particular system is known, the actual counts in the system can be calculated from the observed counts by applying the formula

$$N_0 = \frac{N}{1 - Nt}$$

where N_0 is the actual counting rate, N is the observed counting rate, and t is the dead time. With an average dead time of 250 μ-sec, a loss of 10 percent occurs at counting rates as low as 400 counts/sec, severely limiting the use of the Geiger-Müller counter. However, owing to its ruggedness, the counter is frequently used for crude radiation detection in survey meters.

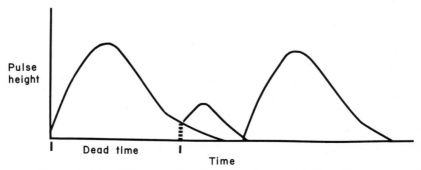

Figure 1–12. Behavior of pulses in a Geiger-Müller detector.

1.4.1.2   Scintillation Detectors

In a scintillation detector (Fig. 1–13) the energy of the nuclear particle is converted into a flash of light in a scintillator. By use of a light pipe and a reflector, much of this scintillator light enters a photomultiplier tube and impinges on a photocathode to produce electrons. The photomultiplier tube amplifies the number of electrons through secondary emission from a series of elements called dynodes. The pulse is multiplied by a factor of approximately 10^6, producing a voltage pulse at the entrance to the preamplifier. Since the amplification factor in the photomultiplier tube is constant, the pulse height is proportional to the energy of the radiation absorbed in the scintillator. The amplitude from the scintillation detector, therefore, resembles that from a proportional detector in that *pulse-height discrimination* is possible; hence one can separate counts from two isotopes by the different energy photons they emit.

The scintillation detector is the most widely used counter in biomedical research. Therefore, it is worthwhile to discuss in some detail the materials used as scintillators. Theoretically, any substance that can undergo luminescence can be used as a scintillator. In this process energy of one type is absorbed and converted into visible radiation. Some common substances used as scintillators are given

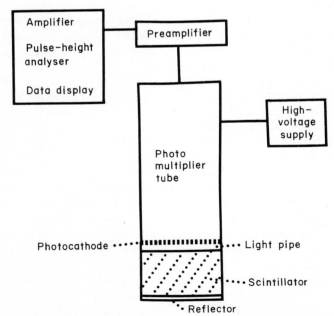

Figure 1–13.   Schematic diagram of a scintillator-type detector.

TABLE 1–1.

Material	Density	Relative Pulse Height for β^- Absorption	Wavelength of Maximum Emission Å
Anthracene crystal	1.25	100	4400
Trans-stilbene	1.15	53	4000
Thallium-activated sodium iodide	3.67	210	4100
Silver-activated zinc sulfide	4.10	210	4500
Toluene containing 4 Gm/l PPO (2.5-diphenyloxazole) and 0.1 Gm/l POPOP (1,4-di(2-(5-phenyloxazolyl)+benzene))	– –	60	4300

in Table 1–1. The organic crystal scintillators are relatively inexpensive but suffer one major disadvantage when compared with the much more expensive inorganic crystal detectors. The inorganic crystals have a greater density and, for comparable crystal size, stop a greater percentage of high-energy gamma radiation. If one wishes to detect gamma radiation, the highest possible efficiency is obtained by using an inorganic crystal.

Liquid Scintillation Counting. Scintillator mixtures dissolved in a solvent (the last item in Table 1–1 is an example) are used to count beta-emitting isotopes. By dissolving all the isotope in the solvent instead of by simply placing a beta-emitting isotope near a crystal, the probability of the beta emission reaching the scintillator is greatly increased.

In Vivo Counting. Gamma radiation has a low absorbance in the human body, and so gamma-emitting radioisotopes can be detected externally by placing a scintillation detector near the skin. Lead collimators (Fig. 1–14) are used to discern the activity from a specific point. Radioactivity in specific regions of the body can be detected by using collimators and inorganic crystal detectors. *Clinical rectilinear scanners attach a moving detector to a "focusing" collimator which detects radiation emanating from a small region (point) in depth at any one moment* (Fig. 1–15). By moving the detector back and forth over the body, the regional radioactivity can be mapped out.

Collimator

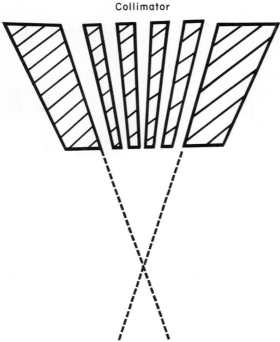

Figure 1–14. A focused collimator; the lead barriers decrease the probability that photons originating from an isotope located outside the focal region will be received by a detector. This effect allows one to detect, with the greatest probability, emissions occurring at one point in space. By changing the length and angle of the holes, the configuration of the iso-response curve can be varied; i.e., iso-response curves connect points with a similar probability of detection by the crystal.

Stationary "cameras" allow for the simultaneous detection of radioactivity from a larger area by using any one of three basic techniques. The most popular effective stationary gamma imaging device is based on the principles described by H. Anger in 1958. The *Anger camera* consists of a large, collimated crystal with 10 photomultiplier tubes to detect the origin of a photon received on the crystal. Another stationary camera, the *Ter-Pogossian camera,* 1963, relies on a very high-gain image-amplifying tube similar to that used in diagnostic radiology, which amplifies the photon detected by a crystal that is incorporated into the image-amplifying tube. The information is displayed on a television screen. The need for a "thin" crystal limits this device to detecting low-energy isotopes. The third stationary imaging device (the *Bender camera,* 1963) uses many small detectors attached in a matrix array and connected by light pipes to an analyzing interface. This detecting system limits the spatial resolution of this instrument.

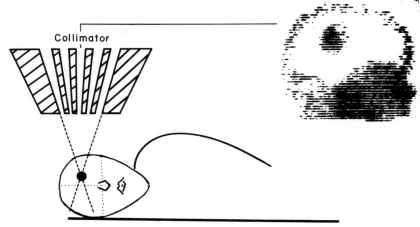

Figure 1–15. A focused collimator similar to that used to detect brain tumors by rectilinear scanning. The collimator "focuses" on a region deep below the surface. Here the collimator crystal combination moves back and forth over the patient to map out the regional radioactivity. This radioactivity is then displayed on a photographic image. Here the brain outline (due to radioactivity within the bone and soft tissue surrounding the brain) and the large brain tumor are seen in the middle of the brain substance, which normally contains virtually no radioactivity. The brain tumor radioactivity can be detected and displayed in this manner.

1.4.1.3   Solid-State Detectors

The solid-state semiconductor detector is similar in principle to the gas-filled ionization detector; however, the charge is carried by electrons and electron vacancies in the semiconductor crystal.

The advantages of the solid-state detector are the following: (1) The use of a thin layer of some semiconductor material having a high density, and hence a high stopping power. (2) The energy required to produce an electron pair in a solid-state detector is approximately one-tenth that required to produce an electron pair in an inert gas. This leads to smaller fluctuations in the number of pairs produced per event, so that much better energy resolution is obtained.

The main disadvantage of the semiconductor detector is that the type commonly available, which uses germanium as the material, has to be operated at the temperature of liquid nitrogen. However, because of the efficiency and count resolution of this type of detector, it will almost certainly find increasing application to biomedical science.

1.5 WHEN ARE TRACERS USED?

Although it seems reasonable that the reader would, at this point, appreciate that tracers are useful, he may not be fully aware of their wide application to problems of biology and medicine. Therefore, having concluded the discussion of the detection and measurement of tracers, it is now appropriate to briefly review where tracers have been used and what can be fruitfully studied by applying tracer principles. In subsequent chapters, selected applications will be discussed in some detail.

In Table 1–2 are listed some of the biomedical applications of tracers, with emphasis on applications of radioactive tracers. Although there are very many tracers and applications which have not been included, this tabulation should give the reader some idea of the important and fruitful areas where tracers have already been applied. This awareness may provide clues to the development of additional applications.

TABLE 1–2. Selected Uses of Tracers

Tracers	System Used	Phenomena Observed
Tritiated thymidine	Autoradiography	Cytokinetics
Barium	Serial X-ray films	Intestinal transit
Diazotrate sodium (intravenous)	Serial X-ray films	Kidney function
Iopanoic acid (telepaque) (oral)	Serial X-ray films	Gall bladder function
Diazotrate meglumin (intra-arterial)	Serial X-ray films	Arterial transit time
Radio transmitter capsule	Radio-receiver	Gastrointestinal acidity
Evans blue dye	Color spectrometer	"plasma volume"
BSP (bromsulphalein)	Color spectrometer	"liver function"
Cardiogreen	Color spectrometer	Cardiac output
Deuterated water (^2HOH)	Mass spectrometer	Body water space
^{14}C-inulin	Liquid scintillation counter	"extracellular water" space
Tritiated water (^3HOH)	Liquid scintillation counter	Body water space
^{131}I-4-iodoantipyrine	Well counter	"body water" space
^{51}Cr-labelled red cells	Well counter	Red cell space
^{131}I-albumin	Well counter	Albumin space (index of plasma volume)
^{125}I-albumin	Well counter	Albumin space (index of plasma volume)
^{24}Na or ^{22}Na	Well counter	Exchangeable sodium
^{42}K	Well counter	Exchangeable potassium
^{82}Br-bromide	Well counter	Extracellular water space
^{131}I-albumin	Rectilinear scanner or Gamma camera	Brain tumor visualization
^{197}Hg & ^{203}Hg chlormerodrin	Rectilinear scanner or Gamma camera	Brain tumor visualization

Table continues on following page

Table 1–2   Selected Uses of Tracers *(Continued)*

Tracers	System Used	Phenomena Observed
$^{99m}TcO_4^-$	Rectilinear scanner or Gamma camera	Brain tumor visualization
^{113m}In-DTPA chelate	Rectilinear scanner or Gamma camera	Brain tumor visualization
^{18}F	Rectilinear scanner or Gamma camera	Bone tumor visualization
^{85}Sr	Rectilinear scanner or Gamma camera	Bone tumor visualization
^{87m}Sr	Rectilinear scanner or Gamma camera	Bone tumor visualization
^{125}I-iodide	Rectilinear scanner or Gamma camera	Thyroid visualization
^{131}I-iodide	Rectilinear scanner or Gamma camera	Thyroid visualization
$^{99m}TcO_4^-$	Rectilinear scanner or Gamma camera	Thyroid visualization
^{131}I-MAA	Rectilinear scanner or Gamma camera	Lung visualization
^{113m}In-labeled particles	Rectilinear scanner or Gamma camera	Lung visualization
^{99m}Tc-MAA	Rectilinear scanner or Gamma camera	Lung visualization
^{99m}Tc-albumin microspheres	Rectilinear scanner or Gamma camera	Lung visualization
^{99m}Tc-sulfur colloid	Rectilinear scanner or Gamma camera	Liver visualization
^{131}I-rose bengal	Rectilinear scanner or Gamma camera	Liver visualization
^{113m}In-colloid	Rectilinear scanner or Gamma camera	Liver visualization
^{198}Au-colloid	Rectilinear scanner or Gamma camera	Liver visualization
^{75}Se-selenomethionine	Rectilinear scanner or Gamma camera	Pancreas visualization
^{203}Hg-chlormerodrin	Rectilinear scanner or Gamma camera	Kidney function
^{131}I-hippuran	Rectilinear scanner or Gamma camera	Kidney function
^{99m}Tc-ascorbic complex	Rectilinear scanner or Gamma camera	Kidney function
^{67}Ga	Rectilinear scanner or Gamma camera	Soft tissue tumor localization
^{67}Ga-citrate	Rectilinear scanner or Gamma camera	Soft tissue tumor localization
^{99m}Tc-TcO_4^-	Rectilinear scanner or Gamma camera	Salivary gland visualization
^{99m}Tc-sulfur colloid	Rectilinear scanner or Gamma camera	Spleen visualization
^{113m}In-colloid	Rectilinear scanner or Gamma camera	Spleen visualization
^{51}Cr-heat damaged red cells	Rectilinear scanner or Gamma camera	Spleen visualization
^{197}Hg-mercurihydroxypropane labeled red cells	Rectilinear scanner or Gamma camera	Spleen visualization
^{99m}Tc-sulfur colloid	Rectilinear scanner or Gamma camera	Bone marrow visualization

Table continues on following page

Table 2–1   Selected Uses of Tracers *(Continued)*

Tracers	System Used	Phenomena Observed
^{113m}In-colloid	Rectilinear scanner or Gamma camera	Bone marrow visualization
^{52}FeCl$_3$	Rectilinear scanner or Gamma camera	Bone marrow visualization
Radioactive Xenon or Krypton	Scintillation probes	Regional blood flow
Radioactive Iodohippuran	Scintillation probes	Renal function studies
^{125}Iodine	Scintillation probes	Thyroid function studies
^{131}Iodine	Scintillation probes	Thyroid function studies
Iron-59	Scintillation probes	Ferrokinetics
^{131}I-HSA	Scintillation probes	CSF dynamics
^{169}Yb-DTPA	Scintillation probes	CSF dynamics

1.6   SOME PROBLEMS ENCOUNTERED IN THE TRACER METHOD

A common problem when using tracers, regardless of type, is the extent to which the tracer approximates the behavior of the tracee. Substitution of one isotope for another of the same element may produce isotopic effects, during which the reaction rates differ owing to mass differences of the isotopes. Deviations from the so-called "ideal tracer" may also occur when the tracer is not a natural constituent of the system (e.g., radioiodine-labeled albumin as a tracer for albumin).

1.6.1   Specific Activity

Specific activity presents another problem when using radio-active tracers. *Specific activity is defined as the ratio of the tracer activity to the amount of tracer.* Tracer activity may be expressed as curies (Ci), counts per minute (counts/min), or disintegrations per second (dis/sec). A curie is the standard measure of activity and equals 3.700×10^{10} dis/sec. The amount of tracer may be expressed as grams, moles, and so forth. Specific activity may, therefore, be expressed as millicuries per gram, and so forth. The use of the term, counts per minute, should be discouraged, because the ratio of counts per disintegrations (efficiency) is a property of the counter and allied electronics, and data expressed in counts per minute cannot be correlated with other data unless the efficiency is given.

The most important quantity to be measured in any tracer experiment is the fraction of the tracer that is contained in a particular organ, tissue, test tube, or whatever. At first glance, specific activity may appear to satisfy this condition, but closer examination needs to be carried out. If a tracer sample is radioactive, how many species (molecules) actually contain the radioactive label? Specific activity does not answer this requirement. Take, for example, tritiated methane CH_3T in which every molecule of the compound contains a tritium atom. The specific activity for such a sample is 42.5 Ci/mmole. On the other hand, methane, in which each molecule is labeled with carbon-14($^{14}CH_4$) has a specific activity of 0.0091 Ci/mmole. *Specific activity is a measure of number of disintegrations, which is related to the half-life (time required for one-half of a sample to disintegrate) and not related to the number of molecules actually labeled.*

1.6.2   Geometric Considerations

When one is counting many samples labeled with an isotope, they must be counted in the same geometric system. If two liquid samples containing the same amount of gamma radiation are counted on a scintillation crystal, relatively fewer gamma rays from the larger sample will be absorbed in the crystal when compared with the smaller sample. Therefore, if many samples are counted in this type of system, they must be of similar volume.

Slight variation in geometry is allowable if one is using a "well counter," where the sample is placed within a hole in a large scintillation crystal. Here a slight variation in a sample size causes only a small variation in absorbed counts. Prior to any experiment it is important to define the limits of geometric variation that will influence the data obtained. This definition of geometric tolerance is usually done by using standard sources or radioactive "phantoms."

1.6.3   Dual or Double Labeling

Important and interesting features of a biomedical system may be studied by labeling an active molecule within the system with two different tags or isotopes. Technically, dual labeling occurs only when the same molecule has two tags or indicators; for example, $^{14}CH_3T$ is a molecule of methane labeled with both carbon 14 and tritium. Double labeling occurs, on the other hand, when a mixture of CH_3T and $^{14}CH_4$ is used as the tracer.

If one attempts to synthesize a species with two labels, usually very few of the molecules actually contain the two labels. For example, methyl mercaptan (CH_3SH), a precursor of methionine, can be prepared by the following reaction:

$$^{14}CH_3I + Na^{35}SH \rightarrow {}^{14}CH_3{}^{35}SH + NaI$$

If the preparation is performed by using specific activities of 1 mCi/mole for each compound, about 5×10^{-5} percent of the actual sulfur is labeled and about 0.1 percent of the carbon. The percent of molecules containing both labels is only $\sim 5 \times 10^{-8}$ percent; the product should, therefore, be written $^{14}CH_3SH + CH_3{}^{35}SH$ rather than $^{14}CH_3{}^{35}SH$. Two labels can be put in the same molecule only:

(1) If the molecule is very large, for example, the iodination of proteins.

(2) If the specific activity of the reactant is extremely high, i.e., carrier free.

1.6.3.1   External Counting of Two Labels

As has been discussed in Section 1.4, gamma-emitting isotopes can be detected in vivo. When using isotopes of widely different energies, the volume of tissue from which each isotope is detected varies due to the different absorptions of the gamma rays in tissue. Thus, if one looks at the two labels simultaneously, he is looking at different anatomic regions; and if one looks at two isotopes with the same detector, he does not see the same total number of counts for the same amount of activity administered. For example, iodide ion labeled with either ^{125}I or ^{131}I will be taken up by the thyroid gland. Using the same detector, one would see a greater percentage of the gamma disintegrations from ^{131}I, because its gamma ray is of considerably higher energy; i.e., more ^{131}I gamma rays would reach the detector, whereas a greater pecentage of ^{125}I gamma rays would be absorbed by the tissue.

It is not easy to compare the anatomic distribution of two labeled compounds at the same time. If two tracers are labeled with the same isotope, they cannot be distinguished separately; if they are labeled with different isotopes, an external detector may have different efficiencies (geometric and intrinsic)[*] for the different energies of the isotopes. It is, therefore, possible to look at two labels only if the gamma rays have similar energies but can still be separated by a spectrometer. Such isotopes are chromium 51 and iodine 131. These isotopes have very similar energies (Fig. 1–16); with a large crystal and fine pulse-height discriminator, they can be separated. With a

---

[*]Geometric efficiency is usually defined as the percent of disintegrations that are received by a detecting system, i.e., $\dfrac{\text{number of photons received by a detector}}{\text{total number of photons emitted}}$, which is influenced only by the geometric relation between the source and the detector. In other words, the source emits radiation in all directions randomly, and the detector usually does not entirely encompass the source of radioactivity. Intrinsic efficiency is usually used to describe the detecting system and is defined as the number of photons striking a detecting system (crystal) divided by the number of photons actually recorded. This is related to the photopeak of the crystal detector and the interaction fraction of the incident events.

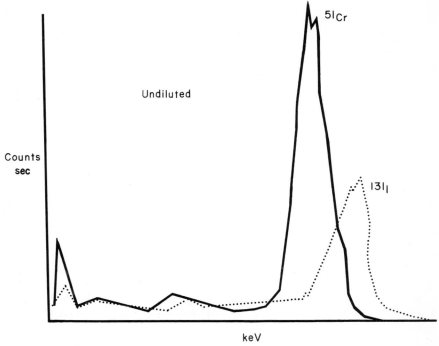

Figure 1–16. Gamma-ray spectrometry used to distinguish ^{131}I from ^{51}Cr. The energy is plotted against relative radioactivity.

3-in. thallium-activated NaI crystal and narrow window the following efficiencies were obtained:

	^{51}Cr efficiency, %	^{131}I efficiency, %
^{51}Cr window	45	8
^{131}I window	0.6	28

Thus, with this system these isotopes could be detected in the presence of one another.

External counting of two labels is useful to compare the amount of albumin (which can be labeled with ^{131}I) and the amount of red blood cells (which can be labeled with ^{51}Cr) in a particular volume shortly after intravenous injection to appraise the regional red cell-albumin ratio, or "regional hematocrit."

The considerations in counting different in vivo isotopes externally resemble the geometry problem associated with static counting. In both cases one has to consider very carefully the geometric efficiency of the counter from the whole sample.

We shall now consider the tools necessary to handle the information obtained from the tracers and detectors. Some, albeit minimal, understanding of mathematical concepts is essential to appreciate how the information can be treated to its best advantage.

BIBLIOGRAPHY

Bell, C. G., and F. N. Hayes (eds.): "Liquid Scintillation Counting," Pergamon Press, New York, 1958.

Kamer, M. D.: "Isotopic Tracers in Biology," 3d ed., Academic Press, Inc., New York, 1957.

Kamer, M. D.: "A Tracer Experiment," Holt, Rinehart and Winston, Inc., New York, 1964.

O'Kelley, G. D.: Detection and Measurement of Nuclear Radiation, *Nat. Acad. Sci. Rep.* NAS-NS-3105, 1962.

Overman, R. T., and H. H. Clark: "Radioisotope Techniques," McGraw-Hill Book Company, New York, 1960.

Price, W. J.: "Nuclear Radiation Detection," 2d ed., McGraw-Hill Book Company, New York, 1964.

Rossi, B. B., and H. H. Staub: "Ionization Chambers and Counters," McGraw-Hill Book Company, New York, 1949.

Chapter 2

SIGNS AND SYMBOLS—AN INTRODUCTION TO MATHEMATICAL CONCEPTS BASIC TO THE TRACER METHOD

Mathematics, both pure and applied, is nothing more than a thought shorthand. This is true whether one is considering the abstract formalism of pure mathematics, which merely tells us that one thing follows another in a certain way (logic), or the concise expression of an idea, condition, or situation in applied mathematics. In this age of the computer, one can well marvel at the effectiveness with which mathematics can deal with very complex problems.

The mathematical representation of an idea or condition is basically in terms of numbers, the most familiar symbolism being, of course, simple arithmetic. As the concepts of algebra* are developed in this chapter, it will be shown that a numerical representation may

---

*Algebra is a generalization of arithmetic, or the use of letters and other symbols to represent numbers. For example, the arithmetic fact $2 + 2 + 2 = 3 \times 2$ is a special case of the general algebraic statement that $x + x + x = 3x$, where x is any number.

be used although actual values may be unknown. Shapes and curves can also be described by algebraic representation.

The authors will present a brief and simple discussion of the two major tools of calculus,* differentiation and integration, which unlock powerful techniques for describing physical changes.

The concepts of matrices† and vectors‡ will also be explored later in this chapter. These mathematical tools are extremely useful when describing conditions and properties that have both magnitude and direction. Any tabulation of data can be more readily stored in a matrix form. Both are also used as information stores.

For example, in studying microvascular transport of plasma proteins these techniques are useful to describe the rate and direction of transport through the capillary endothelium. An appreciation of the techniques used to quantify biologic transport is increasingly important in clinical nuclear medicine.

In fact the application of matrix principles is important to any manipulation of tabular data. The computer storage of time dependent data obtained by a gamma camera requires the use of matrix and vector concepts.

Because all the aforementioned concepts encompass such vast areas, the reader will find, perhaps to his relief, that the authors have attempted to focus on only those tools from arithmetic, algebra, and calculus that are of real and proven use for manipulating data from tracer studies. Naturally, there is bound to be carry over to many other areas in biomedicine requiring mathematical shorthand, particularly any study needing statistical or kinetic analysis.

2.1   ARITHMETIC

2.1.1   Numbers: Their Representation and Operation

At the risk of being trivial, and without attempting a philosophical definition of a number, let us briefly review the kinds of numbers that are most likely encountered. The largest group are termed *real*

---

*Calculus is the field of mathematics which deals with differentiation and integration of functions and related concepts and applications. The concept of a function is discussed in Section 2.2.3.

†A matrix is a rectangular array of terms called elements usually written between parentheses or double lines which is used to work problems in which the relation between the elements is fundamental. An example of a matrix array would be

$$\begin{pmatrix} a_1 & b_1 & c_1 \\ a_2 & b_2 & c_2 \end{pmatrix}$$

‡A vector is a directed line segment in three-dimensional Euclidean space that describes an entity such as force or velocity.

numbers, which are all the positive and negative numbers, including fractions and decimal numbers. Within the realm of real numbers, there are *rational numbers,* which include ordinary fractions and repeating decimals, and *irrational numbers,* which are non-repeating decimals. A subdivision of rational numbers includes the *integers,* which are the whole numbers, both positive and negative, and zero. The positive members of the integers can be referred to as the counting numbers.

Addition, subtraction, multiplication, and division are familiar *operations* in which one uses numbers. These operations obey the basic laws of arithmetic, which are usually taken for granted. These include the following:

$$\text{Commutative law: } a + b = b + a$$
$$a \times b = b \times a$$
$$2 + 3 = 3 + 2$$
$$2 \times 3 = 3 \times 2$$
$$\text{Associative law: } (a + b) + c = a + (b + c)$$
$$(a \times b) \times c = a \times (b \times c)$$
$$(3 + 2) + 1 = 3 + (2 + 1)$$
$$(3 \times 2) \times 1 = 3 \times (2 \times 1)$$
$$\text{Distributive law: } a(b + c) = ab + ac$$
$$3(2 + 1) = 3 \times 2 + 3 \times 1$$

2.1.2 Ratios and Proportions

Two of the more useful concepts in the use of numbers as descriptive tools are the ratio and the proportion. A ratio denotes relative size and may be expressed as a fraction. If there are 3 men and 6 women afflicted with a specific disease, the ratio of men to women is 1 to 2, often written 1:2 or $\frac{1}{2}$; that is, there are half as many men with the disease as there are women.

A proportion is an equality of two ratios or fractions. Such an equality is "1 is to 2" as "3 is to 6," or $\frac{1}{2} = \frac{3}{6}$, or 1:2 = 3:6. That is, proportion is a statement about the relationship of one quantity to another. For example, we may read the above proportion as saying that as some quantity a varies from 1 to 2, then some other quantity b increases from 3 to 6. In other words, quantity a is directly proportional to quantity b. When the first quantity is changed by some ratio, the second changes by the same ratio. If one gram of ^{238}U emits 12,300 alpha particles per second, then two grams emits twice as many, i.e. the decay rate is directly proportional to the weight of the radioactive material.

An inverse relationship is also possible, called an inverse proportion, which means that as one quantity changes by a certain ratio, another quantity changes by the inverse ratio. As a increases from 1 to 2, then b decreases from 2 to 1; or, more generally, as a doubles, b halves. If the pressure of a gas is doubled (and the temperature is maintained constant), the volume of the gas is halved, therefore under these conditions the pressure is evenly proportional to the temperature.

2.1.3 Exponents

If a real number a is multiplied by itself, the product $a \cdot a$ is represented a^2, termed "a square" or "a to the second power." If the product is $a \cdot a \cdot a$, it is represented a^3, termed "a cube," or "a to the third power." It is useful to generalize for the product $a \cdot a \cdot a \cdot a \cdots$ (a multiplied by itself n times) equals a^n, where n is called the exponent and a the base. a^n is a to the nth power. The algebraic notation n is often used to represent the generalized exponent, along with other random letters.

Again, there are certain laws which govern the operations of exponents.

$$\text{Multiplication:} \quad a^n \cdot a^p = a^{n+p}$$
$$2^3 \cdot 2^2 = 2^{3+2} = 2^5$$
$$\text{Division:} \quad a^n/a^p = a^{n-p}$$
$$2^3/2^2 = 2^{3-2} = 2^1$$
$$\text{Exponentiation:} \quad (a^n)^p = a^{n \cdot p}$$
$$(2^3)^2 = 2^{3 \cdot 2} = 2^6$$

The above operations are also true even if the exponent is $-n$ or $1/n$. In the case of $-n$, $a^{-n} = 1/a^n$; that is, a negative exponent indicates a reciprocal. For $1/n$, $a^{1/n} = \sqrt[n]{a}$, that is, the nth root of a. The reader may choose to verify the above laws for operating with $-n$ and $1/n$ exponents by inserting numerical values.

At this point, it is well to note that scientists make extensive use of exponential forms in what is often termed *scientific notation*. For example, take the number 23.257. An equally valid representation is 2.3257×10^1. Or the number 3267.32 may be represented 3.26732×10^3. If a number is less than 1, for example, 0.000123, it may be expressed as 1.23×10^{-4}. The special scientific notation represents any number as a number between 1 and 10 times the number 10 raised to the appropriate power. Suppose a ^{99m}Tc sample decays at the rate of 231,500 disintegrations per minute. In scientific notation the decay rate would be recorded as 2.315×10^5 dis/min.

2.1.4   Logarithms

Let us once again consider the basic representation of a base number raised to a power, $a^b = N$. This expression may also be represented $\log_a N = b$, which is read "The logarithm of N to the base a is b"; that is, *a logarithm is an exponent.* The base number which is most often used is 10; that is, $a = 10$. Consider the following table, wherein is listed the base 10 raised to various powers and the corresponding notation in terms of logarithms.

$$10,000 = 10^4 \qquad \log_{10} 10,000 = \log 10^4 = 4$$
$$1,000 = 10^3 \qquad \log_{10} 1,000 = \log 10^3 = 3$$
$$100 = 10^2 \qquad \log_{10} 100 = \log 10^2 = 2$$
$$10 = 10^1 \qquad \log_{10} 10 = \log 10^1 = 1$$
$$1 = 10^0 \qquad \log_{10} 1 = \log 10^0 = 0$$
$$0.1 = 10^{-1} \qquad \log_{10} 0.1 = \log 10^{-1} = -1$$
$$0.01 = 10^{-2} \qquad \log_{10} 0.01 = \log 10^{-2} = -2$$
$$0.001 = 10^{-3} \qquad \log_{10} 0.001 = \log 10^{-3} = -3$$
$$0.0001 = 10^{-4} \qquad \log_{10} 0.0001 = \log 10^{-4} = -4$$

If shown in scientific notation, all numbers could be represented as numbers from 1 to 10 times 10 raised to the appropriate power.

Now let us show how we may express any number as an exponent of 10. For example, all numbers between 0 and 10 lie between 10^0 and 10^1, numbers from 10 to 1000 lie between 10^1 and 10^3, and so forth. Express the number 2 as an exponent of 10 and a logarithm of 10. Since 2 lies between 10^0 and 10^1, the exponent will be a fraction between 0 and 1; that is, $2 = 10^{0.3010}$, or $\log 2 = 0.3010$.

Some other examples are:

$$17 = 10^{1.2304} \qquad \log 17 = 1.2304$$
$$132 = 10^{2.1206} \qquad \log 132 = 2.1206$$

The laws of operation of logarithms are the same as those for exponents:

Multiplication: $2 \times 17 = 10^{0.3010} \times 10^{1.2304} = 10^{1.5314}$

$\log 2 + \log 17 = 0.3010 + 1.2304 = 1.5314$

Division: $17/2 = 10^{1.2304}/10^{0.3010} = 10^{0.9294}$

$\log 17 - \log 2 = 1.2304 - 0.3010 = 0.9294$

Exponentiation: $2^2 = (10^{0.3010})^2 = 10^{2 \times 0.3010}$

$\log 2^2 = 2 \log 2 = 2 \times 0.3010 = 0.6020$

2.1.5   Natural Logarithms

In the preceding section, it was explained that logarithms are a direct consequence of an exponential expression. It must be noted at this time that although logarithms to the base 10 were specifically illustrated, it is possible to express logarithms in any base number. $a^b = N$ is written $\log_a N = b$, where a is any base number. It is also possible to derive an expression for changing the logarithm of a number expressed in one base to another base:

$$a^b = N \qquad \log_a N = b$$
$$e^c = N \qquad \log_e N = c$$

Since $a^b = e^c$, we can express both sides of the equation in logarithms to the base e:

$$b \log_e a = c \log_e e, \text{ now } \log_e e = 1$$

therefore, $b \log_e a = c$

but $\qquad b = \log_a N$ and $c = \log_e N$

therefore, $\log_e N = (\log_a N)(\log_e a)$

In many scientific expressions it is convenient to express logarithms to a base e, where $e = 2.71828 \cdots$, that is to say, if $e^x = N$, $\log_e N = x$ or $\ln N = x$, where ln denotes logarithm to the base e. These logarithms are called natural logarithms and are also tabulated as are those to the base 10. Using the expression derived above, one can easily convert from one base to the other.

$$\log_{10} e \simeq 0.4343$$
$$\ln 10 \simeq 2.303$$
$$\text{therefore, } \ln N = 2.303 \log N$$

The logarithm of 1 in any base is zero:

$$\log_a 1 = 0 \qquad a^0 = 1$$

Also, the logarithm of

$$\log_a a = 1 \qquad a^1 = a$$

Further, consider again $a^b = N$:

(2.1) $$\log_a N = b$$

Now, raise the first expression to the power m:

$$(a^b)^m = N^m = a^{bm}$$
$$\log_a N^m = bm$$

But if equation (1.1) is multiplied by m,

$$m \log_a N = bm$$

Therefore,

$$\log_a N^m = m \log_a N$$

2.2 FUNDAMENTAL RULES FOR MANIPULATION OF SYMBOLS

2.2.1 Algebra

Algebra is a tool for solving problems in which a quantity under a given set of conditions is unknown. These unknown quantities are designated by letters of the alphabet, and arithmetic operations may be performed upon them as with any number. Algebraic expressions obey all the laws given in Section 2.1.1. In science, the unknown quantity has physical significance; it either is the measurement of a physical property or describes a physical situation.

Algebra provides an additional shorthand notation to describe a particular problem. The unknown quantity is designated by a letter, and equations are devised which describe the set of conditions with which one is concerned.

Example. A vial contains 6 mCi of a radioisotope in 36 cc. What is the volume of 1 mCi? Let x = the volume of 1 mCi. The equation that describes this situation is "$6x = 36$." Therefore, solve for x. $x = 36/6 = 6$ cc.

In algebraic manipulation, the unknown quantity or quantities are known as variables. In a completely general equation the unknown may assume many values compatible with the conditions of the equations.

2.2.2 One-Variable Equations

Let us, first, limit ourselves to equations involving one unknown, or one variable:

Linear Equations.   If a variable appears in an algebraic expression raised only to the first power, it is said to be linear. Similarly, for a linear equation

$$ax + b = 0,$$

the equation is also said to be first degree in x.

Quadratic Equations.   If the variable is raised to the second power, the equation is quadratic, or is second degree in x:

$$ax^2 + bx + c = 0$$
$$ax^2 + c = 0$$

Higher-Degree Equations.   If the equation involves a variable raised to the third power, the equation is termed cubic or third degree. If the variable is raised to the nth power, the equation is one of nth degree.

2.2.3   Functions—Equations With More Than One Variable

Frequently more than one unknown or variable may appear in an equation; specifically, conditions may be such that two or more quantities are unknown; e.g.,

$$y = 3x \quad y \text{ is three times the value of } x$$

Note that any number can be supplied for the value of x and by doing so one has determined the value of y. *When the value of one variable is determined by specifying the value of another variable, the first is said to be a function of the second;* that is, y is a function of x. This may be expressed more generally as the function of x, symbolized $f(x)$, is equal to $3x$. The function is that property which is three times x. If the aforementioned equation in y is rewritten as $f(x) = 3x$, it may also be read as "the function is the value of $f(x)$ at x." At $x = 2$

$$f(x) = 6$$
$$\text{or } f(2) = 3(2) = 6$$

2.2.4   Graph of Functions

As has been noted, the expression $f(x) = 3x$ implies that there are many values of x that one may select for which there is a corres-

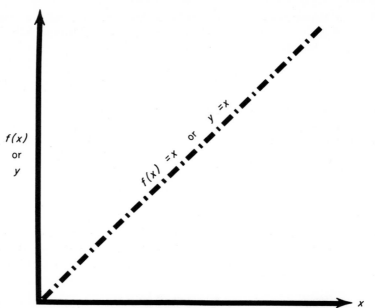

Figure 2–1. Plot of $f(x)$ or y versus x, where $f(x)$ is a linear function of x.

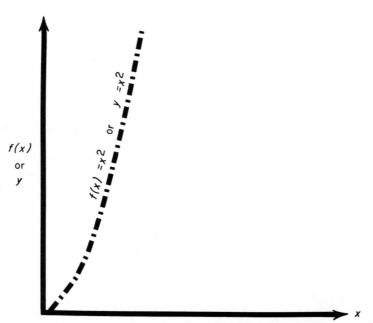

Figure 2–2. Plot of $f(x)$ or y versus x, where $f(x)$ is a nonlinear function of x.

ponding value of $f(x)$. It is, therefore, enlightening to define a function by graphing it. A graph of $f(x)$ versus x serves to define pictorially all values of x and the corresponding values of $f(x)$. See Figures 2–1 and 2–2 for typical examples.

2.2.5   Exponential and Logarithmic Functions

A frequently encountered function in scientific endeavors is an exponential or a logarithmic function

$$f(x) = y = a^x$$
$$f(x) = y = e^x$$
$$\text{or} \quad f(x) = y = \log x$$
$$f(x) = y = \ln x$$

Plots of some exponential and logarithmic functions are shown in Figures 2–3 and 2–4.

Figure 2–5 is a plot of $N = N_0 e^{-\lambda t}$. This exponential equation is of particular interest because it describes radioactive decay or spontaneous disintegration. N_0 is the number of radioactive nuclei at some

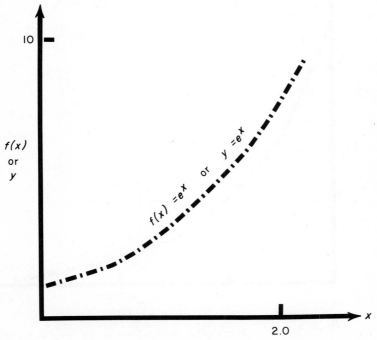

Figure 2–3.   Plot of the exponential function e^x versus x.

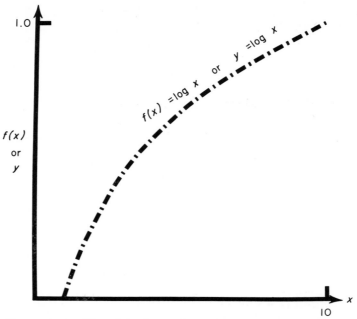

Figure 2–4. Plot of the logarithmic function log *x* versus *x*.

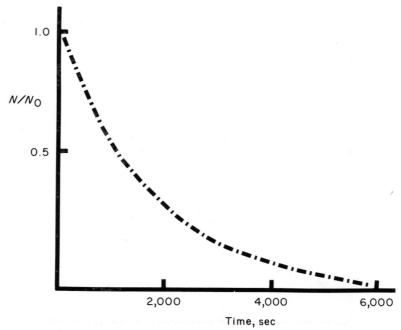

Figure 2–5. Typical exponential radioactive decay for 1224-sec half-life carbon 11.

time $t = 0$; N is the number of radioactive nuclei remaining after time t; λ is a constant called the decay constant, characteristic for each isotope. The curve in Figure 2–5 is often referred to as an exponential decay curve, and the example given is for carbon-11, an isotope of carbon with a 1224-sec half-life.

The decay constant, λ, for ^{11}C is 5.56×10^{-4} sec^{-1}. If one has N_0 nuclei of ^{11}C initially, after t sec $N = N_0 e^{-\lambda t}$ nuclei of ^{11}C. If we rearrange the equation to $N/N_0 = e^{-\lambda t}$ and let $N/N_0 = 1/2$, that is, N is one-half of N_0 or one-half the original number of nuclei, then $t = t_{1/2}$, the time required for one-half the number of nuclei to disintegrate; $t_{1/2}$, or the half-life, is a characteristic property of all radioactive isotopes.

2.3   ANALYSIS

2.3.1   Dimensions

The use of mathematics as a scientific tool is dependent upon the interpretation of the mathematical expression in physical terms. An equation, though perhaps mathematically pleasing, is of little interest to an applied scientist unless it describes a physical situation or change. The concept of physical units or dimensions for variables (parameters) in a mathematical expression is extremely important. Dimensions undergo the same operations as the parameters that they describe. If two parameters are multiplied, their corresponding physical units are multiplied also; a similar rule holds for quotients.

Example 1.   The area of a rectangle is the product of its length times its width, $A = lw$. For a rectangle 2 cm long and 1 cm wide, the area $= (2 \text{ cm})(1 \text{ cm}) = 2 \text{ cm}^2$.

Example 2.   Take the formula $R = PV/T$, which is the ideal gas law. What are the units of R? P is pressure, V is volume, and T is temperature. If we express P in atmospheres, V in liters, and T in

degrees, then $R = \dfrac{\text{atmospheres} \quad \text{liters}}{\text{degrees}}$.

The choice of units is arbitrary or governed by the convention that specifies the size of a basic unit. The numerical magnitude of a parameter is arbitrary and will, of course, change if the basic units are changed. In Example 2, just discussed, if one had expressed P in millimeters and V in cubic centimeters, the units of R would have

been $\dfrac{\text{cubic centimeters} \quad \text{millimeters}}{\text{degrees}}$, where the actual numerical value of R would be quite different from its value expressed as $\dfrac{liter\ atmosphere}{degree}$. If a parameter is determined by a quotient in which numerator and denominator have the same units, the resulting quantity is dimensionless. Its numerical value will be independent of the units that are chosen.

Dimensional analysis of a physical problem is extremely important and can direct the scientist to the correct arithmetic or algebraic operations that must be performed in order to obtain the desired answer.

2.3.2   Digital Versus Analog Concepts

Step Functions Versus Continuous Functions. Our general definition of a function did not delineate between a step function (Fig. 2–6) and a continuous function (Fig. 2–7). A step function is discontinuous over a specified range, hence its name. These functions are visual examples of the kinds of data that are generated in many tracer studies. The step function represents digital data, or the measurement of some physical property at various intervals. For example, a physician observing his patient at various intervals is generating

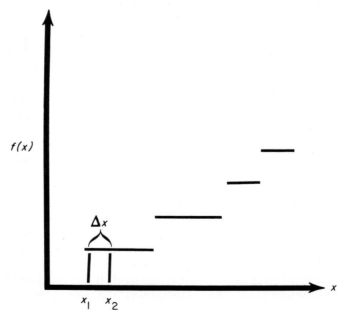

Figure 2–6.   Plot of $f(x)$ versus x for a typical step, or discontinuous, function.

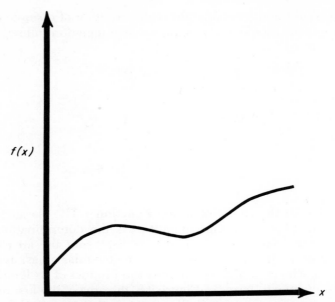

Figure 2-7.   Plot of $f(x)$ versus x for a typical continuous function.

digital data. The continuous function represents analog data, or measurement of a property continuously over an interval. If the physician were to watch his patient continuously 24 hours a day, he would be collecting analog data.

Step Functions.   Let us examine the properties of step functions. Step functions can be considered as precise, constant functions. They are constant over at least one interval. The intervals of a step function are discrete or definite increments. The interval from x_1 to x_2 may be represented as $\Delta x = x_2 - x_1$, where the symbol delta Δ denotes a measurable change from x_1 to x_2. We often want to represent the addition of many intervals or increments, i.e.,

$$\Delta x_1 + \Delta x_2 + \Delta x_3 + \cdots + x_n \quad \text{where } \Delta x_1 = x_2 - x_1, \ldots$$

This summation may be represented in a shorthand notation:

$$\sum_{n=1}^{n} = \Delta x_n$$

which is read "the summation of all Δx between x_1 and x_n." The summation symbol is the capital Greek letter sigma (Σ); the numbers (or letters) on the bottom and top of the sigma tell the range over which we are summing. The bottom number is the lower limit and the top the upper limit. This summation notation is perfectly general. It must be emphasized that a summation is the addition of discrete measurable parameters. Some other examples of summation follow:

$$\sum_{n=1}^{3} n^2 = 1^2 + 2^2 + 3^2$$

$$\sum_{k=1}^{4} X_k = X_1 + X_2 + X_3 + X_4$$

$$\sum_{n=1}^{5} n = 1 + 2 + 3 + 4 + 5$$

Area Under the Curve of a Step Function. The classic problem of the summation of various quantities is the determination of the area under the curve. It is possible to represent the area under a step function as the summation of the rectangular areas, as seen in Figure 2–8. The area of a rectangle is the product of its length times its width. Written in terms of Figure 2.8, the area of the first rectangle is $f(x_1)(x_2 - x_1)$ or, more generally, $f(x)(\Delta x)$; that is, $f(x)$, the value of the step function (length), times Δx, the value of the interval from

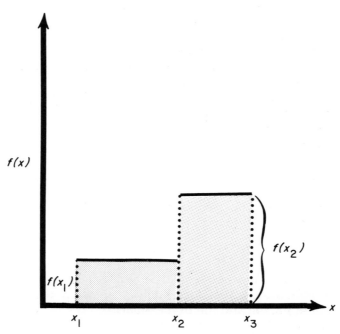

Figure 2–8.   Area under a step function.

x_1 to x_2 (width). The total area can then be represented as the summation of all the individual areas. Area equals

$$\sum_{n=0}^{n} f(x_n)(\Delta x_n).$$

It is well to emphasize again that the summation Σ is used when the intervals either in a step function or in some data-collecting experiment are discrete, definite, measurable quantities. Experimentally, this means we have measured some property at certain time intervals, e.g., checked a patient's blood pressure every two hours. In this case, the units of x will be time, and Δx will be equal to 2 hr.

Continuous Functions.   Let us consider now the properties of continuous functions. We can, indeed, extend the idea of a step function to the continuous function. We may look upon our continuous function as a kind of step function in which the intervals are so infinitesimally small that the interval, in fact, becomes a point. This infinitesimally small interval is represented dx, and the summation of all dx may be represented as

$$\int_{x_1}^{x_n} dx,$$

where $\int$ is the summation of all dx from x_1 to x_n. The symbol $\int$ is an integral sign. The numbers or letters below and above indicate the range over which the integration (summation) is performed. Again, the top number is the upper limit and the bottom number the lower limit.

Area Under the Curve of a Continuous Function.   What about the area under a continuous curve? The analog of the step function also holds; that is, the area is the summation of all the rectangles that can be constructed of infinitesimally small width. As shown in Figure 2–9, the area of an infinitesimal rectangle is $f(x)[(x + dx) - x]$ or $f(x)\ dx$, and the total area equals

$$\int_{x_1}^{x_n} f(x)\ dx$$

The concept of an integral as the area under a curve gives an intuitive and pictorial handle in the understanding of the integration operation. The aforementioned summation is the general expression of an integral. The example of the integral in the preceding section is, in fact, a special case of $\int f(x)\ dx$. In $\int dx$, $f(x)$ equals 1, and $\int dx$ is the area under the constant function $f(x) = 1$. The integral lends itself then to summation of changing quantities.

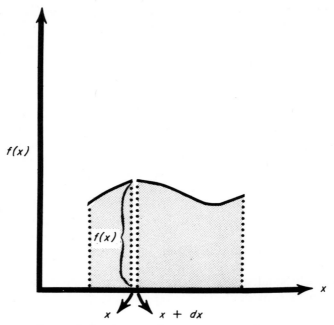

Figure 2–9. Area under a continuous function.

The actual evaluation of this area or integral is often a very complicated operation. It can, in fact, be evaluated through the summation operation we have described, but the task is tedious. After the subsequent discussion on differentiation, one will see how the task of finding integrals may be simplified. Let it suffice for now to present some integration methods for simple functions without any proof of the method.

Example 1. Consider the integral of x^k, where k is any power of x except -1 evaluated between a and b:

$$\int_a^b x^k \, dx = \frac{b^{k+1} - a^{k+1}}{k+1}$$

$$\int_2^3 x^2 \, dx = \frac{3^{2+1} - 2^{2+1}}{2+1} = \frac{3^3 - 2^3}{3} = \frac{19}{3}$$

For $k = -1$, we have the special case illustrated in the following. We shall pursue this case in a later section.

$$\int_a^b x^{-1} \, dx = \ln b - \ln a$$

Example 2. Any constant times a function may be placed outside the integral sign:

$$\int_1^3 3x^2 \, dx = 3 \int_1^3 x^2 \, dx = 3 \, \frac{3^3 - 1^3}{3} = 26$$

More generally,

$$\int_a^b cf(x) \, dx = c \int_a^b f(x) \, dx$$

Example 3. Polynomials may be integrated term by term:

$$\int_1^2 (x^2 + 5x - 3) \, dx = \int_1^2 x^2 \, dx + 5 \int_1^2 x \, dx - 3 \int_1^2 dx$$

$$= \frac{2^3 - 1^3}{3} + 5 \, \frac{2^2 - 1^2}{2} - 3 \, \frac{2 - 1}{1}$$

$$= \frac{7}{3} + \frac{15}{2} - 3 = \frac{41}{6}$$

More generally,

$$\int_a^b [c_1 f(x) + c_2 g(x)] \, dx = c_1 \int_a^b f(x) \, dx + c_2 \int_a^b g(x) \, dx$$

Some other properties of integrals are:

$$\int_a^b f(x) \, dx + \int_b^c f(x) \, dx = \int_a^c f(x) \, dx$$

Thus, we have

$$\int_1^2 x^2 \, dx + \int_2^3 x^2 \, dx = \int_1^3 x^2 \, dx$$

This is merely the summation of two individual areas to give a total area:

$$\int_a^b f(x) \, dx = \int_{a+c}^{b+c} f(x - c) \, dx$$

For example,

$$\int_1^2 x \, dx = \int_{1+3}^{2+3} (x - 3) \, dx = \int_4^5 (x - 3) \, dx$$

$$\int_2^4 (x - 1)^3 \, dx = \int_{2-1}^{4-1} x^3 \, dx = \int_1^3 x^3 \, dx$$

Often, changing the limits in the above example simplifies the evaluation of the integral:

$$\int_a^b f(x)\,dx = -\int_b^a f(x)\,dx$$

$$\int_1^3 x^4\,dx = -\int_3^1 x^4\,dx$$

In other words, interchanging the upper and lower limits gives the integral a negative sign.

Let us also include the integral for $f(x) = e^x$,

$$\int_a^b e^x\,dx = e^b - e^a$$

because tracer methods often involve exponential functions. Tables of integration are provided in most calculus books and mathematics handbooks.

2.3.3   Differentiation — Rate of Change

In many experiments, one is faced with the question, "How fast is a certain quantity changing?". The answer to this question can be found by differentiation—the other major concept of calculus. The rate of change is the slope of the line connecting $f(x_1)$ and $f(x_2)$ (or y_1 and y_2) (Fig. 2–10). This line, or any line joining two points on a curve, is a secant line, and its slope is given by

$$\frac{y_2 - y_1}{x_2 - x_1} = \frac{\Delta y}{\Delta x}$$

(that is, the change in y divided by the change in x, where $y_2 = y_1 + \Delta y$ and $x_2 = x_1 + \Delta x$)

$$= \frac{[f(x_1 + \Delta x) - f(x_1)]}{\Delta x}$$

Now as the intervals between x_1 and x_2 become smaller and smaller, the rate of change, in fact, becomes the slope of the line tangent to the curve at that point.

$$\text{Slope of the tangent} = \frac{dy}{dx}$$

or, more formally,

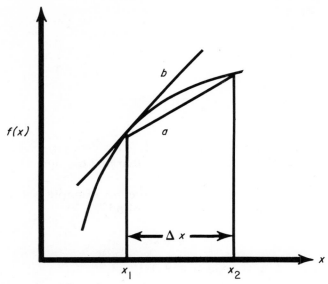

Figure 2-10. Plot of $f(x)$ versus x for a typical continuous function, where line a is the secant line whose slope is $\Delta y/\Delta x$ and line b is the tangent line whose slope is dy/dx as $\Delta x \to 0$.

$$\lim_{\Delta x \to 0} \frac{\Delta y}{\Delta x} = \lim_{\Delta x \to 0} \frac{f(x_1 + \Delta x) - f(x_1)}{\Delta x} = \frac{dy}{dx}$$

This terminology, lim $\Delta y/\Delta x$, means that as Δx approaches zero, or becomes smaller and smaller, $\Delta y/\Delta x$ equals dy/dx and is called the derivative of y with respect to x. The notation $f'(x)$ is also used for derivative; in this case, it is the derivative of $f(x)$ with respect to x.

Again, a few limited examples of the differentiation operation are presented here. More extensive tables may be found in calculus books or mathematics handbooks.

Find the derivative of a straight line, the general equation for which is $y = mx + b$, where m is the coefficient of x, and b is a constant. It is possible to find the derivative by the difference-quotient method described previously:

(2.2) $y = mx + b$

(2.3) $y + \Delta y = m(x + \Delta x) + b = mx + m\,\Delta x + b$

Subtract (2.2) from (2.3):

$$\Delta y = m\,\Delta x$$

Divide by Δx:

$$\frac{\Delta y}{\Delta x} = m$$

Now

$$\frac{dy}{dx} = \lim_{\Delta x \to 0} \frac{\Delta y}{\Delta x} = m$$

We may point out that a constant function $y = b$ is a special case of the general case for the straight line. It is clear that the derivative of $y = b$ is zero.

How about the derivative of a curve, for example, $y = x^2$? Again, let us apply the difference-quotient method:

$$(2.4) \qquad y = x^2$$

$$(2.5) \qquad y + \Delta y = (x + \Delta x)^2$$

$$= x^2 + 2x\,\Delta x + \Delta x^2$$

Subtract (2.4) from (2.5):

$$\Delta y = 2x\,\Delta x + \Delta x^2$$

Divide by Δx:

$$\frac{\Delta y}{\Delta x} = 2x + \Delta x$$

$$\frac{dy}{dx} = \lim_{\Delta x \to 0} \frac{\Delta y}{\Delta x} = 2x$$

Let us explore one more case, $y = x^3$:

$$(2.6) \qquad y = x^3$$

$$y + \Delta y = (x + \Delta x)^3$$

$$= x^3 + 3x^2\,\Delta x + 3x\,\Delta x^2 + \Delta x^3$$

$$\Delta y = 3x^2\,\Delta x + 3x\,\Delta x^2 + \Delta x^3$$

$$\frac{\Delta y}{\Delta x} = 3x^2 + 3x\,\Delta x + \Delta x^2$$

$$\frac{dy}{dx} = \lim_{\Delta x \to 0} \frac{\Delta y}{\Delta x} = 3x^2$$

We may now generalize the differentiation operation on curves of the form $y = x^n$:

$$\frac{dy}{dx} = nx^{n-1}.$$

Let us now return to an important function, the logarithmic function, and attempt to find the derivative of this curve. It shall be seen later that this function and its integral and differential properties are extremely important in the discussion of tracer kinetics.

$$y = \log x$$

$$y + \Delta y = \log (x + \Delta x)$$

$$\Delta y = \log (x + \Delta x) - \log x$$

$$\Delta y = \log \frac{x + \Delta x}{x}$$

$$= \log \left(1 + \frac{\Delta x}{x}\right)$$

Divide by Δx:

$$\frac{\Delta y}{\Delta x} = \frac{1}{\Delta x} \log \left(1 + \frac{\Delta x}{x}\right)$$

If one attempts to proceed with the limit operation, it is found that $1/\Delta x$ goes to zero and $\log (1 + \Delta x/x)$ goes to $\log 1$, which is zero. So one is left with the curious result of $0/0$; surely, by intuition, the slope of a tangent line to a logarithmic curve is not zero. Let us make a substitution at this point, $\Delta x/x = t$. Therefore $\Delta x = xt$:

$$\frac{\Delta y}{\Delta x} = \frac{1}{xt} \log (1 + t)$$

From our rules of logarithms this may be rewritten

$$\frac{\Delta y}{\Delta x} = \frac{1}{x} \log (1 + t)^{1/t}$$

When the limit operation is applied, one is faced with the problem of evaluating

$$\lim_{t \to 0} \log (1 + t)^{1/t}$$

Remember $\Delta x/x$ goes to zero as $\Delta x \to 0$; therefore, as $\Delta x \to 0$, $t \to 0$.

If one calculates $(1 + t)^{1/t}$ as t is made smaller and smaller, e.g., for $t = 1/10.000$, $(1 + t)^{1/t} = 2.7182$. Even as t is made smaller and smaller, the limit, in fact, becomes $2.7183 \cdots$. It will be recalled from Section 2.1.5 that the peculiar number $e = 2.7183 \cdots$. This then is the definition of e:

$$e \equiv \lim_{t \to 0} (1 + t)^{1/t}$$

where the sign $\equiv$ means identically equal to. In other words, by definition

$$\frac{\Delta y}{\Delta x} = \frac{1}{x} \log e \quad \text{in any base.}$$

If the logarithmic base number is taken as e, it will be recalled that $\ln e = 1$; therefore,

$$\frac{dy}{dx} = \lim_{\Delta x \to 0} \frac{\Delta y}{\Delta x} = \frac{1}{x}$$

For $y = \ln x$,

$$\frac{dy}{dx} = \frac{1}{x}$$

Included without proof is the derivative of the function $y = e^x$:

$$\frac{dy}{dx} = e^x.$$

The authors have indicated some of the more elementary functions and their derivatives. Let us list some rules for operating with derivatives. The verification of these results may be found in any calculus text.

If
$$y = cf(x), \text{ then } \frac{dy}{dx} = cf'(x)$$

where c represents any constant.

If
$$y = f(x) + g(x), \text{ then } \frac{dy}{dx} = f'(x) + g'(x)$$

If
$$y = f(x) = g(x), \text{ then } \frac{dy}{dx} = f'(x) = g'(x)$$

If
$$y = f(x)g(x), \text{ then } \frac{dy}{dx} = f(x)g'(x) + g(x)f'(x)$$

If
$$y = \frac{f(x)}{g(x)}, \text{ then } \frac{dy}{dx} = \frac{g(x)f'(x) - f(x)g'(x)}{[g(x)]^2}$$

2.3.4   Relationship Between Differentiation and Integration

The reader should about now be considering the question, "Is there a connection between integration and differentiation?" The actual answer is that they are analogous to the processes involved in squaring a (positive) number and then taking the square root of the squared number. One of course returns to the original number. In terms of functions, the relationship means that if a function is integrated, a new function results. If the new function is, in turn, differentiated, one returns to the original function.

The authors have presented the integral as an area; now it is logical to ask at what rate this area changes.

$$A(x) = \int_a^x f(x)\,dx$$

$A(x)$ is the area equal to the integral of some function $f(x)$.

Let us consider the function $f(x) = x^2$, and perform integration:

$$\int_a^x x^2\,dx = \frac{x^3 - a^3}{3}$$

If we now differentiate the function $x^3/3 - a^3/3$, we have

$$f(x) = \frac{x^3}{3} - \frac{a^3}{3}$$

$$f'(x) = \frac{3x^2}{3} - 0$$

$$= x^2.$$

In more general terms of area,

$$A(x) = \int_a^x f(x)\,dx$$

If the area has changed by the dimensions of the rectangle given in Figure 2–11, where x has increased to $x + dx$, then dA, the change in area, is equal to $f(x)\,dx$:

$$(A + dA) - A = f(x)\,[(x + dx) - x]$$

$$dA = f(x)\,dx$$

Divide by dx:

$$\frac{dA}{dx} = f(x)$$

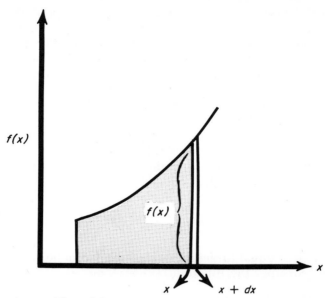

Figure 2–11. Plot of $f(x)$ versus x for a continuous function where the change in the area under the curve equals the area of the next infinitesimal rectangle.

Therefore,

$$\frac{dA}{dx} = \frac{d \int f(x)\, dx}{dx} = f(x)$$

or

$$A'(x) = f(x)$$

The operation of integrating a function that is a derivative of another function is often called antidifferentiation, or finding the antiderivative or primitive. From the previous example, we can see that any two primitives of the same function will differ by a constant.

If $f(x) = 2x$, since dy/dx of $x^2 = 2x$, an antiderivative or primitive of $2x$ is x^2. Another primitive of $2x$ is $x^2 + 2$ or $x^2 + 3$; that is, the primitive of $2x$ must be represented as $x^2 + c$, where c is an arbitrary constant that may be evaluated by the specific conditions of the problem. Therefore, the following equations may be written:

$$\int kx^n\, dx = \frac{kx^{n+1}}{n+1} + C \quad n = -1$$

$$\int kx^{-1}\, dx = \int k\, \frac{dx}{x} = k \ln x + C \ .$$

$$\int e^{kx}\, dx = \frac{1}{k}\, e^{kx} + C$$

It is now appropriate to apply the tools of calculus to physical problems. Before proceeding, it is necessary to say a word about differential equations. Differential equations involve at least one derivative of an unknown; e.g.,

$$\frac{dy}{dx} = \sin x$$

$$\frac{d^2y}{dx^2} + k^2y = 0$$

$$\frac{dS}{dt} = 0.04S$$

Some of the preceding equations must look very familiar. In order to solve a differential equation, one should find a relationship connecting the variables that satisfies the differential equation. Essentially, this is the operation of antidifferentiation, by which one can find the function that gave rise to the derivative. Solutions to differential equations may be simple, as with our examples of finding primitives, or may be very complex and involve making certain approximations in order to solve the equations. The study of differential equations involves a wide spectrum of events and problems, and we have again limited our use of differential equations to those problems peculiar to tracer methods.

Let us now consider a problem that was used earlier without full explanation, the problem of radioactive decay. The rate of disintegration of a radioisotope is proportional to the amount present at time t. If N represents the number of radioactive atoms present at time t, then the rate of disintegration is $-dN/dt$; in other words, the change of N (that is, dN) with respect to t. Notice the negative sign for the derivative, because it represents a decreasing rate of change, which is, in turn, proportional to N.

$$\frac{-dN}{dt} = kN$$

We may separate our variables,

$$\frac{dN}{N} = -k\,dt$$

and integrate both sides of the differential equation. The result is

$$\ln N = -kt + C$$

or

$$\ln N = -kt + \ln c$$

Since C is an arbitrary constant, it is as valid to use $\ln c$ as it is to use C. Now at $t = 0$, $N = N_0$ is the number of atoms originally present, or $c = N_0$. Therefore,

$$\ln N = -kt + \ln N_0$$

or

$$N = N_0 e^{-kt}$$

The reader will recall that this is exactly the result used as an example in an earlier section (Sec. 2.2.5).

The above solution to the differential equation explicitly defines the condition of the physical situation and is exactly the solution expected when interpreting first-order kinetic problems, as discussed in Chapter 6.

The reader has seen that the two major branches of calculus are interrelated, and that this relationship is most useful in describing specific physical problems encountered when applying the tracer method.

2.4 MATRICES AND VECTORS

Preceding sections have briefly introduced the notation and operations of ordinary algebra. The notation and operations of matrices, which are here presented, are merely a later development in the shorthand of ordinary algebra; in a sense, a still shorter hand or a more abbreviated notation.

Matrix algebra arises from the necessity to solve simultaneous linear equations and to transform the linear equations from one set of variables to another. In many situations in mathematics one is confronted with equations of linear transformation. Take for example two simultaneous equations in two unknowns:

$$a_1 x + b_1 y = h_1$$
$$a_2 x + b_2 y = h_2$$

The unknowns in these equations may easily be transformed from the rectangular coordinates used to another coordinate system, such as polar coordinates.

One need only imagine the tedium involved in the manipulation of a set of m simultaneous equations in n unknowns to appreciate the need for matrix algebra. The theory of matrices, however, though part of algebra, possesses a utility beyond the domain of ordinary algebra.

Matrix algebra has been found to be the best means to express many ideas of applied mathematics. For example, its most powerful application was in 1925 when Heisenberg introduced matrices into the mathematics of quantum mechanics. Today matrices are omnipresent in both pure and applied mathematics, being an especially powerful tool in solving many problems in medical and biological sciences. Whenever there are many interlocking relationships, matrices appear on the scene and facilitate our understanding of the processes.

2.4.1   Matrix Notation

The following abbreviated notation for a set of simultaneous equations was presented in the last century (Cayley, 1857):

$$\begin{bmatrix} a_{11} & a_{12} & \cdots & a_{1n} \\ a_{21} & a_{22} & \cdots & a_{2n} \\ & & \cdots & \\ a_{m1} & a_{m2} & \cdots & a_{mn} \end{bmatrix} \begin{bmatrix} x_1 \\ x_2 \\ \cdot \\ \cdot \\ \cdot \\ x_n \end{bmatrix} = \begin{bmatrix} y_1 \\ y_2 \\ \cdot \\ \cdot \\ \cdot \\ y_n \end{bmatrix}$$

that is, a detached rectangular scheme of the coefficients a_{ij} from the variable x_j. These detached rectangular schemes are matrices. The detached-coefficient scheme is referred to as an operator acting upon the variables $x_1, x_2, \ldots, x_n$ in much the same way as a acts upon x to produce ax. The rules of operations of these schemes are matrix algebra.

A matrix, quite simply, is a rectangular array, scheme, or table of numbers set out in m rows and n columns. Such a matrix is referred to as a matrix of order m by n or $m \times n$. The numbers in the array or scheme are called elements and may be generally represented as a_{ij}, where a_{ij} is the element in the ith row and jth column. The matrix as a whole is represented by a capital letter such as $A = [a_{ij}]$ or $B = [b_{ij}]$, the brackets implying all values of a in i rows and j columns. It must be noted that each matrix is a complete entity. If, for example, we interchange any of its rows, or its columns, we, in general, obtain a different matrix. Two matrices A and B are considered equal only when they are of the same order $m \times n$ and when all corresponding elements agree, i.e., when $a_{ij} = b_{ij}$ for all ij.

Up to now, elements of matrices have been presented as pure numbers or mathematical expressions. But in practical application, the elements of matrices come from the external world and have physical, economic, or social meaning. There are numerous ways in which matrices may be used as carriers of information. Information storage and tabulation are one of the most important uses of matrices in biomedical sciences. Consider, for example, the following tabulation of incidence of respiratory disorder and habitat expressed as a fraction of the population with the designated complaint in each locale.

	Persistent Cough	Chest Illness	Sputum Volume 2 ml. or More
Cities	0.334	0.107	0.350
Suburban	0.242	0.069	0.272
Rural	0.215	0.062	0.228

Each row records the incidence of all complaints in the locale. Each column records the fraction of people with the complaint in each locale. The reader is reminded that each day newspapers (especially the sports page, which contains the win-loss records of each team against each team) are filled with such tabulations or matrices. Thus, the use of matrices as data-storage devices is quite familiar.

2.4.2   Vectors

Before proceeding with the rules of matrix operation, consider the special case of a matrix which consists of a single row, or a single column. Such matrices are very common and are called vectors; more precisely, row vectors or column vectors. Vectors are $1 \times m$ or $n \times 1$ matrices.

At this point, it is perhaps helpful to digress to an alternative description of vectors. Vectors may also be defined geometrically. Consider blood flowing at a certain velocity in a particular direction which could be represented by a vector. A vector represents a property having both magnitude and direction; a vector quantity may be represented graphically as an arrow, the length of which is representative of the magnitude and the head of which is pointed in the specified direction. Figure 2–12 shows such a representation. Also shown is a vector operation that easily relates geometrical development of vectors to the matrix algebra development of vectors.

A vector V in Figure 2–12 may be resolved into components, in this case two components — one in the x direction and one in the y direction. In other words, vector V may be considered as composed of vector c_1 in the y direction and vector c_2 in the x direction. The components, as indicated, are simply found by drawing a perpendicular line from the head of vector V to each axis. A vector with two components is a two-dimensional vector; a vector with n components is an n-dimensional vector.

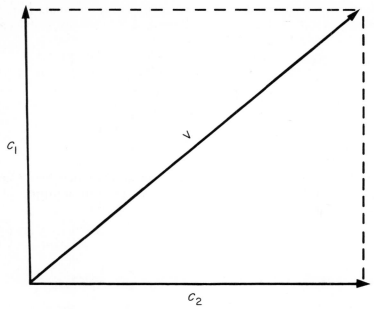

Figure 2–12. Resolution of a vector into two components.

It should be obvious that a vector may be defined by specifying its components. Our matrix notation is well suited for this kind of specification of a vector. The components of a vector may be arranged in either a row or a column, each number (element) being the magnitude of a component of the vector. The total number of elements is the dimensionality of the vector. For example, $V = [2 \quad 3 \quad 5]$ is a vector in three dimensions, where 2, 3, 5 are the coefficients of some basis, or unit vector, against which we are measuring other vectors. If **i**, **j**, **k** are the unit vectors, then $V = 2i + 3j + 5k$ (boldface type represents vector quantities).

2.4.3 Operational Rules of Matrices

The rules of matrix operation (and vector operation) are as few and simple as those of ordinary algebra. The commutative law, associative law, identity operation, and inverses (analogs of reciprocals) also operate in matrix algebra. Although it is beyond the scope of this text to present a detailed description of matrix theory and operations, the authors will offer some examples of matrix operation and indicate potential application to the tracer method and biomedical sciences.

Matrix Addition. Two matrices A and B of the same order $m \times n$

may be added by summing their corresponding elements to form a sum matrix $A + B$. Consider the following 2×3 general matrices:

$$A = \begin{bmatrix} a_{11} & a_{12} & a_{13} \\ a_{21} & a_{22} & a_{23} \end{bmatrix} \quad B = \begin{bmatrix} b_{11} & b_{12} & b_{13} \\ b_{21} & b_{22} & b_{23} \end{bmatrix}$$

The addition of $A + B$ is

$$A + B = \begin{bmatrix} a_{11} + b_{11} & a_{12} + b_{12} & a_{13} + b_{13} \\ a_{21} + b_{21} & a_{22} + b_{22} & a_{23} + b_{23} \end{bmatrix}$$

Matrix addition is commutative, that is, for any two matrices A and B that can be added, $A + B = B + A$. Let us perform addition of the matrices

$$\begin{bmatrix} 1 & 3 & -2 \\ -1 & 4 & 0 \end{bmatrix} + \begin{bmatrix} 5 & 0 & 1 \\ 1 & -2 & 3 \end{bmatrix} = \begin{bmatrix} 1+5 & 3+0 & -2+1 \\ -1+1 & 4-2 & 0+3 \end{bmatrix}$$

$$= \begin{bmatrix} 6 & 3 & -1 \\ 0 & 2 & 3 \end{bmatrix}$$

or the addition of the vectors

$$[2 \quad 5 \quad -3] + [7 \quad 4 \quad 2] = [9 \quad 9 \quad -1]$$

It is well to recall that element-by-element addition is also common in ordinary algebra. For example, add

$$z = 3x - 4y$$

to

$$2z = 2x + 3y$$

The sum $(1+2)z = (3+2)x + (-4+3)y$, is obtained by adding the corresponding coefficients.

Matrix Subtraction. Two matrices A and B of the same order $m \times n$ may be subtracted by subtracting their corresponding elements. For example,

$$\begin{bmatrix} 4 & -1 \\ 2 & 3 \end{bmatrix} - \begin{bmatrix} 3 & 7 \\ 5 & -x \end{bmatrix} = \begin{bmatrix} 1 & -8 \\ -3 & 3+x \end{bmatrix}$$

Matrix Multiplication. A matrix A of order $m \times n$ may be multiplied by a matrix B of order $n \times p$ to give a product matrix AB of order $m \times p$. The following example illustrates matrix multiplication:

$$\begin{bmatrix} 1 & 3 & -4 \\ 2 & 5 & 2 \end{bmatrix} \times \begin{bmatrix} 3 & 1 \\ 2 & 4 \\ 5 & -1 \end{bmatrix}$$

$$= \begin{bmatrix} (1)(3)+(3)(2)+(-4)(5) & (1)(1)+(3)(4)+(-4)(-1) \\ (2)(3)+(5)(2)+(2)(5) & (2)(1)+(5)(4)+(2)(-1) \end{bmatrix}$$

$$= \begin{bmatrix} -11 & 17 \\ 26 & 20 \end{bmatrix}$$

that is, the ijth element of the product matrix is formed by the sum of the products of corresponding elements of the ith row of matrix A and the jth column of matrix B. We may express this in the more general form for the multiplication of a 2×2 matrix by a 2×2 matrix:

$$A = \begin{bmatrix} a_{11} & a_{12} \\ a_{21} & a_{22} \end{bmatrix} \qquad B = \begin{bmatrix} b_{11} & b_{12} \\ b_{21} & b_{22} \end{bmatrix}$$

$$AB = \begin{bmatrix} a_{11}b_{11} + a_{12}b_{21} & a_{11}b_{12} + a_{12}b_{22} \\ a_{21}b_{11} + a_{22}b_{21} & a_{21}b_{12} + a_{22}b_{22} \end{bmatrix}$$

We now come upon an important distinction between matrix multiplication and ordinary algebraic multiplication: matrix multiplication is not commutative; that is, $AB \neq BA$. Let us illustrate by writing out the general expression of the multiplication of B by A:

$$BA = \begin{bmatrix} b_{11}a_{11} + b_{12}a_{21} & b_{11}a_{12} + b_{12}a_{22} \\ b_{21}a_{11} + b_{22}a_{21} & b_{21}a_{12} + b_{22}a_{22} \end{bmatrix}$$

We can clearly see that all elements of the product matrix AB are different from those of the product matrix BA.

Matrix Division. In ordinary algebra, the process of division is that of solving the equation $ax = b$ for the unknown x. The solution is expressed as

$$x = \frac{b}{a} \qquad \text{or } x = a^{-1}b \qquad \text{or } x = ba^{-1}$$

If, on the other hand, one considers the matrix equation

$$AX = B$$

the solution for the unknown matrix X is somewhat more complicated than the solution in ordinary algebra, for matrix multiplication is not commutative. Therefore, the solution of X in the above equation can be different from the solution of Y of the equation

$$YA = B$$

The problem of matrix division is soluble if we introduce two more concepts of matrix algebra which are actually suggested by ordinary algebra. If a matrix A has a reciprocal matrix A^{-1}, then $A^{-1}A = I$, where I is the identity matrix, analogous to the number 1 in ordinary algebra. The identity matrix is simply a matrix in which all the diagonal elements are 1 and all other elements are zero. A 3×3 identity matrix is

$$\begin{bmatrix} 1 & 0 & 0 \\ 0 & 1 & 0 \\ 0 & 0 & 1 \end{bmatrix}$$

The reader may verify that the multiplication of any matrix times the identity matrix yields the original matrix; that is, $AI = A$. Let us multiply both sides of the equation by A^{-1}:

$$A^{-1}AX = A^{-1}B$$

but

$$A^{-1}A = I$$

Therefore,

$$IX = A^{-1}B$$

or

$$X = A^{-1}B$$

Matrix division is then the problem of finding the reciprocal matrix of A and multiplying by B.

Consider the following example:

$$\begin{bmatrix} 0 & 2 \\ 2 & 0 \end{bmatrix} X = \begin{bmatrix} 1 & 2 \\ 3 & 4 \end{bmatrix}$$

The reciprocal matrix of A is

$$\begin{bmatrix} 0 & \frac{1}{2} \\ \frac{1}{2} & 0 \end{bmatrix}$$

Therefore

$$X = \begin{bmatrix} 0 & \frac{1}{2} \\ \frac{1}{2} & 0 \end{bmatrix} \begin{bmatrix} 1 & 2 \\ 3 & 4 \end{bmatrix} = \begin{bmatrix} \frac{3}{2} & 2 \\ \frac{1}{2} & 1 \end{bmatrix}$$

Matrix division, however, is complicated by two aspects: (1) the actual calculation of the reciprocal matrix; and (2) the fact that some matrices have no reciprocals. The answers to these complications are not simple and shall not be covered here. However, they may be found by the reader in any matrix algebra text. The authors have merely attempted to outline some of the operations that one may form with matrices and to point out their similarities to and differences from ordinary algebra.

2.4.4   Example of Use of Matrices

Before ending the discussion of matrices, let us again turn to the example dealing with the incidence of respiratory disease and habitat. In the matrix illustrated previously, incidence of respiratory disease and locale was tabulated. Let us now assume that this study was carried out in three different areas of the country, each of which may be subdivided into city, suburban, and rural locales. The following matrix tabulates these data, with the elements of the matrix being the populations of the locales:

Area	City	Suburban	Rural
1	520,320	120,120	6,230
2	20,320	320,330	130,010
3	0	100,530	373,000

If we now make use of the technique of matrix multiplication, multiplying the matrix of complaint versus habitat by the above matrix, the product matrix tabulates the total incidence of the three respiratory complaints in the three areas studied.

Area	Persistent Cough	Chest Illness	Sputum Volume 2 ml or More
1	204,770	64,288	225,970
2	111,350	32,335	123,260
3	104,750	30,030	112,350

When the clinical data are available, matrix algebra will be very helpful in determining the usefulness of various scanning agents. For example, let us consider three scanning agents A, B, and C; their efficiency in diagnosing various abnormal states can be represented in matrix form.

	X	Y	Z
A	0.90	0.85	0.60
B	0.85	0.70	0.95
C	0.80	0.90	0.75

The matrix shows the probability of diagnosing diseased states X, Y, and Z with the three agents. A one-dimensional matrix can be used to express the relative occurrence of the states X, Y, and Z in the subjects scanned.

	X	Y	Z
Occurrence of state (relative to $X=1$)	1.00	0.50	0.30

By multiplication of the two matrices we arrive at a third matrix, which is a one-dimensional matrix giving the relative diagnostic efficiency of the three agents.

Relative Diagnostic
Efficiency

A	1.505
B	1.485
C	1.500

These are but limited examples of the use of matrices as storers of information and of their manipulability through operational rules similar to ordinary algebra to give us new information. Matrix algebra is a powerful tool for solving many problems in biomedical science; its use has grown tremendously in recent years. This section has attempted to survey its simple operational rules and to suggest its potential for describing physical situations.

BIBLIOGRAPHY

Aitken, A. C.: "Determinants and Matrices," 9th ed., Interscience Publishers, Inc., New York, 1956.

Davis, P. J.: "The Mathematics of Matrices," Blaisdell Publishing Company, New York, 1965.

Greenberg, D. A.: "Mathematics for Introductory Science Courses," W. A. Benjamin, Inc., New York, 1965.

Kruglak, H., and J. T. Moore: "Basic Mathematics for the Physical Sciences," McGraw-Hill Book Company, New York, 1963.

Schwartz, J. T.: "Introduction to Matrices and Vectors," McGraw-Hill Book Company, New York, 1961.

Chapter 3

STATISTICAL METHODS

The preceding chapter dealt with some tools of mathematics that may be used to describe physical phenomena and their respective changes. The application of tracer methods also requires the collection, organization, analysis, and interpretation of data. Statistics is the branch of applied mathematics that deals with such data processing; more precisely, with the mathematical characterizations of aggregates (observations, measurements, or items). Statistical application extends to all disciplines; not only the physical and biological sciences, but also the social and political sciences.

For tracer, as well as other biological experiments, one trial or run is usually not satisfactory; one does not ordinarily express confidence in a single measurement. Most experiments are designed such that the measurement is repeated several times; each time it is repeated, the numerical value usually differs from previous trials. In repeated trials, one is asking the question; "Is the observation reproducible?". Can a "reliable" or "reproducible" value be obtained? It has already been indicated, however, that this reliable or reproducible value is not a single number. Instead, a distribution of measurements about some average or mean value is obtained.

The methods in the following sections will deal with the means to obtain "reliable values" from tracer experiments. Statistical methods will be used to (1) analyze data; (2) determine errors in a set of data; (3) show correlation (or lack of it) between sets of data; (4) determine the best curve that may be drawn through sets of data points.

59

3.1 ANALYSIS OF DATA

Let us begin with a discussion of some basic concepts of statistics, then proceed with their application to the specific requirements in tracer experiments.

3.1.1 Probability

The cornerstone of statistics is the concept of probability. Comments like "There is an even chance that...," or "It is likely that...," are simply statements of probability that a certain event will occur. The actual outcome is uncertain, but there is some degree of confidence that an event will occur. Statistics provide us with the mathematical framework for dealing with statements of probability.

Probability (P) is expressed as the ratio between the number of times a certain result will occur in repeated trials and the total number of such trials in an experiment. If P is equal to 1, then the event is certain to occur, and one can predict with certainty the result from an experiment. If P equals 0, the event will not occur or a certain result from an experiment will not be obtained. From practical experience, it is known that scientific experiments do not usually lie at these extremes. Normally one expresses that the probability is high that a certain result will be obtained or is low that a certain result will be obtained. If a probability of 0.9 is assigned to a certain result, it is meant that 90 per cent of the time this result will be obtained if the experiment is repeated indefinitely. The previous statement actually suggests two other basic problems of statistics: (1) How does one assign such a measure to experimental efforts; i.e., how is it known that if the experiment is repeated indefinitely, in 0.9 (90 per cent) of the trials one shall obtain the certain result. This is a problem of distributions, which involves not only the probability that a certain result will be obtained, but also the probabilities of obtaining other results. (2) The basic problem of statistics comes when one reconsiders the statement "repeated indefinitely." Obviously an experiment cannot be repeated indefinitely. How many times must one repeat an experiment to obtain a "reliable" number? This is the statistical problem of sampling, the problem of how large a set of results must be to represent the actual outcome of an experiment. As the reader shall see, certain kinds of distributions impose certain sample sizes.

For now, let us simply define a sample as a set of points representing the possible outcomes of an experiment. Probability may be expressed in terms of frequency. The probability assigned to a given sample point A should be equal to the proportion of times the point will be obtained in a large number of repeated experiments. The probability of A occurring is the number of points that represent A occurring divided by the total number of points obtained.

$$P(A) = \frac{\text{number of outcomes representing } A \text{ occurring}}{\text{total number of outcomes obtained in the experiment}}$$

$$= \frac{n(A)}{n}.$$

The probability is simply related to the frequency with which A occurs.

The calcuation of probability for a single event is quite simple and may seem intuitive. For example, consider the toss of a common six-faced die. What is the probability of turning up a 4? The total number of outcomes for a single toss of the die is six. Therefore, the probability of turning up a 4 is

$$P = \frac{1}{6}$$

What is the probability of turning up an even number? There are three outcomes (2, 4, 6) which represent an even number turning up. The probability is

$$P = \frac{3}{6} = \frac{1}{2}$$

(In these examples, it has been assumed that the die is not loaded and that the faces are *equally likely* to turn up.)

In most experiments, however, the assignment of probability for a certain result is not so simply made. Probability assignment must come from compilation of data and experimental estimation.

3.1.2   Distributions—Patterns of Variation

In the preceding section, the authors have indicated the close relationship between the probability that an event will occur and the related frequency of the event's occurrence in a large number of trials. Attention is now directed to determining the frequency of an event's occurrence. Raw data, a set of measurements, are usually in an unsuitable form to obtain significant information. The experimenter may note the smallest and largest values and that the other values somehow scatter between these extremes. Clearly a sum-

TABLE 3–1.

Reaction Rate — Unorganized Data

20.7	23.3	26.1	26.7	24.3
29.8	23.5	23.1	27.3	22.3
24.6	24.2	23.7	28.4	25.4
25.8	24.9	21.5	21.9	25.3
24.3	27.8	22.6	22.5	24.8

marized, systematic tabulation would aid interpretation of the experimental points.

The simplest organization which one may impose on the original data is to collect alike values and to arrange them in numerical order. This order forms a *frequency distribution* of the original data, i.e., a generalized tabulation of the variation of the experimental points. It is usually necessary to group data into appropriate classes before their general characteristics may be detected. This process of classification is simply the collection of all experimental points which lie within a certain interval or class. The end points of the interval are called class boundaries, and the midpoint of upper and lower class boundaries is called the class midpoint or mark.

Let us consider a specific set of data to illustrate classification and distribution. The data in Table 3–1 were obtained in measuring the rate of a certain biochemical reaction. The *frequency* in Table 3–2 is simply the number of points which lie within a given class interval. The *relative frequency* is simply the frequency divided by the total number of measurements.

The process of classifying the data is equivalent to a coarser rounding off of the original data. There are no rigorous sets of rules for determining the length of class intervals, but for most data 10 to 20

TABLE 3–2.

Frequency Table of Data in Table 3–1

Class	Class Boundaries	Class Midpoints	Frequency	Relative Frequency
1	20.0–21.0	20.5	1	0.04
2	21.0–22.0	21.5	2	0.08
3	22.0–23.0	22.5	3	0.12
4	23.0–24.0	23.5	4	0.16
5	24.0–25.0	24.5	6	0.24
6	25.0–26.0	25.5	3	0.12
7	26.0–27.0	26.5	2	0.08
8	27.0–28.0	27.5	2	0.08
9	28.0–29.0	28.5	1	0.04
10	29.0–30.0	29.5	1	0.04
Total			25	1.00

classes are desirable. The length of the class interval would, therefore, be about one-tenth to one-twentieth of the difference between
the largest and smallest measurement. For the data in Table 3–1,
this difference is $29.8 - 20.2 = 9.6$; for 10 intervals, the length should
be about 1. The set of frequencies with its respective classes is called
the *frequency distribution.*

Graphical representations of frequency distributions are useful
in their study. The most common and widely used graph of a frequency distribution is a histogram. A histogram is a vertical bar
graph constructed by marking off the class intervals on the horizontal
axis and frequency units on the vertical axis. For each class interval, a
rectangle is constructed, whose base is the length of the interval and
whose height is the frequency of that class.

Figure 3–1 is a histogram of the data in Table 3–2. In constructing
a histogram, one must keep in mind that the areas of the rectangles
should be proportional to the corresponding frequencies. If for some
reason class intervals of unequal sizes are used, the rectangle heights
are calculated so that frequency is proportional to the area in a histogram; hence, if some rectangles have double the width of others, the
altitudes of the former must correspond to half the frequencies of
the larger class.

In addition to the frequency scale in Figure 3.1 we have included
a relative-frequency scale. If the overall area of a histogram represents
frequency of occurrence and the area of each rectangle is proportional
to the frequency of each respective class, then the area is also proportional to relative frequency, which in turn is related to the probability
of obtaining a value in the class interval in question.

A somewhat different construction of a histogram is often useful

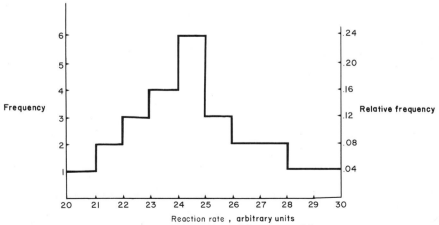

Figure 3–1. Histogram of data in Table 3–2.

for comparison purposes. If one wishes to see the probability of contracting a particular disease at various ages, the total area of the histogram is made equal to the probability of contracting the disease during the total lifetime and the area of each rectangle is the probability of contracting it during a particular age group.

Distribution Functions. To recapitulate, a distribution may be used as a measure of the probability that a certain property, quantity, or event being measured will assume a certain value. If one envisions the property being measured as a variable, it is possible to think of functions describing the probability that the variable will assume a particular value. This is much the same as our study of functions between x and y in the preceding chapter. Such a function is called a *distribution function* and is a shorthand notation for the probability that a variable x will assume certain values.

From our study of functions, it is known that they may be discrete or continuous and that their variables may be discrete or continuous. The same is true for distribution functions that may be constructed in statistics. If a variable can assume only specific values (usually integers), it is called a discrete or discontinuous variable. Examples of discrete variables are a specific number of children in a family or a specific number of radioactive disintegrations. If a variable can assume any value whatsoever between certain limits, it is called a continuous variable. Examples of continuous variables are any measurements of length, weight, and time between two limits. A discontinuous distribution function is shown in Figure 3–2. A continuous distribution is simply represented as a smooth curve.

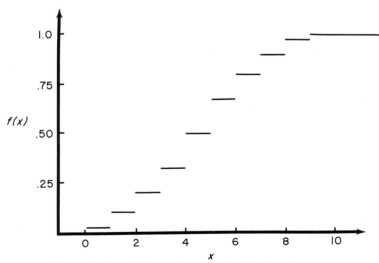

Figure 3–2. A discontinuous distribution function.

As in calculus, it is much easier to deal with continuous curves than with step functions. For example, let us consider an experiment in which is measured the thyroid ^{131}I uptake 6 hr after oral administration in male adults from a given city. Although there will be only a finite number of measurements made and therefore a finite number of possible outcomes, the calculations are simplified if we assume an infinite number of individuals and introduce x, the uptake, as a continuous variable (i.e., x may assume any value in certain limits); a continuous curve or distribution function is thereby formed. We might also consider a discrete function as approaching a continuous curve; as smaller and smaller intervals are chosen, the step function will smooth out and approximate a continuous curve.

Figures 3–3 to 3–5 show examples of frequently encountered distribution functions. Figure 3–3 is a binomial distribution, which is used for an experiment of the repetitive type in which only occurrence or nonoccurrence of an event is recorded. Figure 3–4 is a Poisson distribution, which shall be discussed further in a later section dealing with its application to counting of radioactive materials. Figure 3–5 is a normal distribution to which a large class of physical measurements conforms. It should be noted that the normal distribution is very similar in shape to the binomial and Poisson distributions.

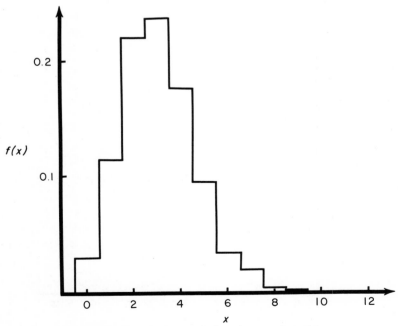

Figure 3–3. Binomial distribution function.

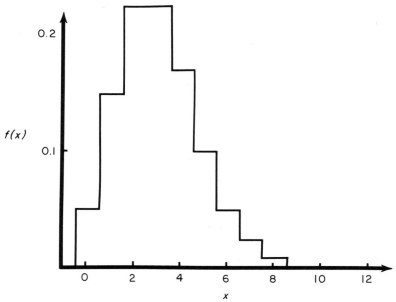

Figure 3–4.   Poisson distribution function.

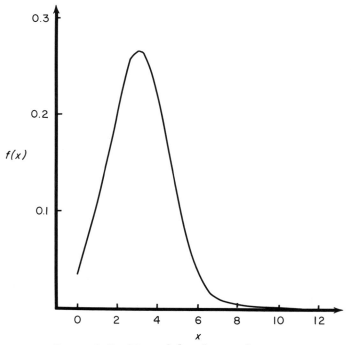

Figure 3–5.   Normal distribution function.

3.1.3   Measures of Central Tendency

In studying distribution functions, it is useful to further simplify a mass of data by defining certain measures that describe important features of the distribution. One is a measure of the center of the distribution, or central tendency; another is the scatter of the data about this central point. The latter will be considered in the next section.

Arithmetic Mean.   The most familiar and most frequently used measure of central tendency is the *arithmetic mean,* or mean, or what is usually called the "average." *The mean is defined as the sum of all the values of the measurement or observations divided by the total number of measurements or observations,* or, expressed in terms of our summation-sign (Σ) shorthand,

$$\overline{x} = \frac{x_1 + x_2 + x_3 + \cdots + x_n}{n} = \frac{\sum\limits_{i=1}^{n} x_i}{n}$$

where $\overline{x}$ is the mean, and x_i are all the values of x that have been observed. If we have made 10 observations with the values of 1, 2, 3, 3, 4, 4, 5, 6, 6, 7, then we have

$$\overline{x} = \frac{\sum\limits_{i=1}^{10} x_i}{10} = \frac{1+2+3+3+4+4+5+6+6+7}{10} = \frac{41}{10} = 4.1$$

The arithmetic mean is easy to compute, easy to define, takes all measurements into consideration, and is well designed for algebraic manipulation.

Median.   Another measure of central tendency is the *median,* which is the middle value when the numbers compressing the data are arranged in ascending or descending magnitude. For example, for a set of measurements 2, 2, 3, 4, 4, 5, 5, 5, 6 the median is 4. Quite simply, *the median is that point on the scale of measurement above which and below which 50 per cent of the numbers fall.* The median is not readily affected by extreme measurements. This stability makes it a useful measure in many cases where the extreme values of the variable are considered due to unusual circumstances. Although less frequently used than the arithmetic mean, the median is often useful when an accurate determination of the arithmetic mean is impossible.

Geometric Mean.   The *geometric mean* is less frequently used than the arithmetic mean or median. The geometric mean is defined as

$$M = \sqrt[n]{x_1 x_2 x_3 \cdots x_n}$$

It is interesting to note that the logarithm of the geometric mean is equal to the arithmetic mean of the logarithms. The geometric mean is applied to data that form a geometric progression.

3.1.4 Measures of Dispersion

In addition to a measure for studying the centering of a set of data, it is desirable to have a measure of the spread, or dispersion, the variability of the data about the center. Are the data closely bunched about the center or widely spread?

Range. The simplest measure of dispersion is the *range*, which is defined as the difference between the largest and the smallest value in the set of data.

$$R = \text{largest value} - \text{smallest value}$$

Although the range does give some indication of the spread of data about the center, it depends solely upon extreme values and not at all on the way in which other values fall between the extremes. The range is particularly useful for sets of data where n is less than 10, because as n increases there is a greater chance that one will obtain an extreme value. However, it is definitely a measure of the "width" of the data.

Mean Deviation. Another measure of dispersion is the *mean deviation*, which is defined as the absolute sum of the differences between the observed values in a set of data and the mean divided by the total number of values. The notation of *absolute value* is, perhaps, not at once apparent. It may be easily shown that the sum of the differences between the observed values and the mean is zero. This is, of course, of no value as a measure of dispersion. However, if one considers that it is immaterial whether a value deviates positively or negatively about the mean, the sign of the deviation may be ignored, and the average of the deviations (all taken as positive) measures the dispersion about the mean:

$$MD = \frac{\sum_{i=1}^{n} |x_i - \bar{x}|}{n}$$

Although the mean deviation is easy to define and compute, its main disadvantage is that quantities involving computation of absolute value are unsuitable for algebraic manipulations.

Another method for using the deviations from the mean and preventing cancellation is squaring the deviations. Squaring makes quantities positive. The average of the squared deviation from the mean is called the *variance*. For statistical reasons, which shall not be discussed, the *variance* is more properly defined as

$$S^2 = \frac{\sum\limits_{i=1}^{n} (x_i - \bar{x})^2}{n - 1}$$

Except in the case of small samples, the choice between n and $n-1$ makes little difference in our results.

The units of the variance are the square of the original units. It is often convenient, however, to express a measure of variation in the same units as the original data. The square root of the variance is such a measure and is called the *standard deviation*

$$s = \sqrt{s^2} = \sqrt{\frac{\sum\limits_{i=1}^{n} (x_i - \bar{x})^2}{n - 1}}$$

It is not easy to grasp at once the significance of the standard deviation, but it has important properties with respect to the normal distribution. A normal distribution is defined as one that has 68 per cent of the measurements in a set of data lying between the interval $\bar{x} - s$ and $\bar{x} + s$, and the interval $\bar{x} - 2s$ and $\bar{x} + 2s$ includes 95 per cent of the measurements. The mean and standard deviation, therefore, completely determine a normal distribution (Fig. 3–6).

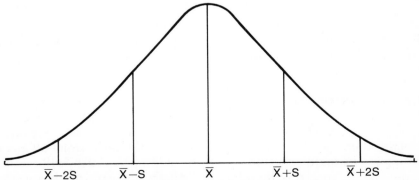

Figure 3–6. A normal distribution showing the mean and standard deviation. Within the limits $\bar{x} - s$ and $\bar{x} + s$ lie 68 per cent of the observed data; within the interval $\bar{x} - 2s$ and $\bar{x} + 2s$ lie 95 per cent of the data.

3.1.5 Sampling

The totality of possible experimental outcomes in any experimental situation is called the *population* of the outcome. The set of data obtained from performing the experiment a number of times is called the *sample*. We are here concerned with generalizing the properties of the sample of the population; i.e., if an experiment is performed n times, the properties (distribution, mean deviation, etc.) of the sample (n) are the same as those of a sample $2n$ in size, and also the same as those of the total population.

Sampling is one of the most important concepts in statistics. It poses the central problem of how big must the sample be to contain all the information about the population. The answer to this question is given in two steps: First, how large an error can be tolerated in the estimate? Second, what confidence limits must be expressed for the allowable error? In the following section on errors, we shall discuss these points in detail. Let it suffice, at this point, to say that for a normal distribution over which a large class of physical measurements conforms, it is generally necessary to have about 20 measurements.

In addition to the problem of sample size, one can well visualize the situation where there is more than one sample from a population. We then have more than one sample mean and may ask how these different sample means distribute themselves. The answer is quite simple; the distribution of the means is a normal distribution, even though the original population is not normally distributed. It is often useful to express a standard deviation for all the means from the various samples. This standard deviation of the means is called the *standard error* of the means and is defined as

$$s_x = \frac{s}{\sqrt{n}}$$

The standard error gives a measurement of the limits within which one would expect the mean of the next sample to lie.

3.2 DETERMINATION OF ERRORS

It is practically impossible to measure any physical quantity exactly; there is always some uncertainty in the measurement. Even the most elaborate experiments, if repeated many times, will yield a variety of answers. It is, therefore, obligatory in any physical measurement to specify the amount by which the result may be in error. Such errors will now be discussed.

3.2.1   Experimental Errors

Experimental errors may be classified as *systematic (determinate)* and *random (indeterminate)*. Systematic errors are caused by a defect in the experimental method itself or by malfunctions of the apparatus used to perform the experiment. These errors often may be large but may usually be minimized by a modified plan of attack. Systematic errors may remain relatively constant and affect all measurements in the same way. They appear to be independent of experimental conditions. On the other hand, these errors may vary systematically with one or more experimental conditions. It is generally possible to detect determinate errors by a series of carefully controlled experiments in which experimental conditions and quantities are varied widely in a systematic way.

Random errors are unavoidable because there is some uncertainty in every physical measurement. Even the observation of discrete real events will be distorted by their very measurement. This distortion appears as random errors. Examples of random errors are instrument drift or variations in scale reading by the observer. These errors are usually small deviations from the correct value and are as likely to be positive as negative. The statistical methods we have been considering deal with these errors.

Of particular interest is the random decay of radioisotopes. If we place a radioactive material in a counter, the disintegration rate, the fraction of radiation impinging on the counter, and the counting rate are all subject to random fluctuations. All scientific data must be reported with the error involved in the measurement; for example, a counting rate may be given as 4000 ± 60 counts/min.

3.2.2   Combination Errors

Suppose that our desired unknown quantity is a combination of several measurable quantities, each expressed with an error; how does one determine errors in derived quantities? For example, assume a count of a radioactive sample in a counter. The sample counts $10,200 \pm 101$ counts/min. The background is 200 ± 14 counts/min. The net counting rate is $(10,200 \pm 101) - (200 \pm 14)$; what is the error in the net counting rate? If we either sum or subtract two quantities to obtain a third, the maximum error in the third is the sum of the absolute values of the errors in the two quantities summed or subtracted. Therefore, the net counting rate is $10,000 \pm 115$ counts/min.

If, on the other hand, two quantities are to be multiplied to obtain a third, the relative errors (or percentage errors) add to give the relative error in the derived quantity. For example, multiply 100 ± 10 times 30 ± 3; this may be expressed in terms of relative or percentage errors:

$$(100 \pm 10\%)(30 \pm 10\%) = 3000 \pm 20\% = 3000 \pm 600$$

The error is obtained the same way for a quotient.

3.2.3   Confidence Limits

It is usually impractical to make enough measurements for the sample size to even approach the size of the population. One must, therefore, be satisfied with a sample of sufficient size to calculate sample parameters rather than population parameters. For example, it must suffice to calculate $\bar{x}$, the sample mean, instead of μ, the true population mean. Even with a few determinations, however, it is possible to predict within what limits the sample mean is likely to agree with μ. That is, one can express a certain probability that the true mean lies within some range of the sample mean. This range is designated as the *confidence limit*. For example, one may specify a range of values about the sample mean and state that there is a probability of 0.5 (50 per cent) or a 50-50 chance that the true value will lie within the range. The usual confidence limits used in studies are 95 or 99 per cent probability.

We may use the "student's t" distribution to set the confidence limits. The use of a t calculation is based on probability and the number of degrees of freedom of the sample. The *degree of freedom* of a sample are the number of measurements in the sample minus one $(n-1)$. The quantity t is defined as

$$t = \frac{\bar{x} - \mu}{s/\sqrt{n}}$$

which means that t is the deviation between the sample mean and the true mean measured in units of the standard error, $s/\sqrt{n}$. If several t's were calculated from samples of a population, the t's would distribute according to the so-called t distribution, a distribution symmetrical about the mean. The t distribution is particularly geared to small samples. For large samples, the distribution approaches a normal distribution.

The confidence limits for $\bar{x}$ using t are

$$\bar{x} \pm \frac{t\,s}{\sqrt{n}}$$

For a certain probability (or risk, the opposite of probability) and degrees of freedom, t is constant and is found in a table of t values.

Example. Let us take a set of nine measurements for which we want to find the 95 per cent confidence limits of the sample mean. The mean of this sample is 2.55 and $s = 0.05$; for $N = 9$ or degrees of freedom $(n-1)$ equal 8 and a risk α of 0.05 (if $P = 0.95$, then $\alpha = 0.05$) t is 2.31. Therefore,

$$\bar{x} \pm \frac{t\,s}{\sqrt{n}} = 2.55 \pm \frac{(2.31)(0.05)}{\sqrt{9}} = 2.55 \pm 0.04$$

One can predict with 95 per cent confidence that the true mean will lie between 2.59 and 2.51.

3.2.4   Significance Tests

One is often faced with the problem of evaluating or comparing data statistically. For example, suppose that there are two samples of n measurements of a certain property; are the respective means of the two samples significantly different? Significance testing involves setting up a hypothesis and then testing it. Our hypothesis for the example at hand would be that the means of the two samples are the same. Such a hypothesis is called a null hypothesis. Our task is then to determine whether or not the measurements in the samples offer evidence against or contradict the null hypothesis. A test of significance yields a probability. If the probability associated with the test is small, then one concludes either that an exceptional event has occurred or that the hypothesis is not correct.

In testing the hypothesis, one must decide what to call a small probability. This value, called the significance level, is usually taken equal to or less than 0.05. If the test of significance results in a probability of 0.05 or less, assuming the null hypothesis to be true, the outcome obtained would occur 5 times or less in 100 events. When this occurs, we reject the null-hypothesis test.

The t distribution affords us a simple test of significance. Assume two samples of means $\bar{x}_1$ and $\bar{x}_2$. Are they significantly different? The null hypothesis will be that the means of samples are the same, that is, that $\mu_1 = \mu_2$. We now define t in terms of the following formula:

$$t = \frac{(\bar{x}_1 - \bar{x}_2) - (\mu_1 - \mu_2)}{\sqrt{s_1^2/n_1 + s_2^2/n_2}}$$

If $\mu_1 = \mu_2$, then $\mu_1 - \mu_2 = 0$ and t is

$$t = \frac{\bar{x}_1 - \bar{x}_2}{\sqrt{s_1^2/n_1 + s_2^2/n_2}}$$

If the calculated t is less than the t value in a t table at the appropriate confidence level, then the hypothesis is accepted and there is no significant difference between the two samples. If the calculated t is greater than the table t value, the hypothesis is rejected and the difference is significant.

Another frequently used test of significance is the *chi-square test*. This test is used to determine whether observed values of some property are significantly different from the assumed or expected value. The chi-square test gives a basis for deciding whether the differences can be caused by random error. Chi-square is defined as

$$\chi^2 = \frac{\sum\limits_{i=1}^{R} (o_i - e_i)^2}{e_i}$$

where o_i is the observed value and e_i is the expected or assumed value, provided there is some expected value at hand. Frequently, however, no such expected or theoretical value is available, in which case chi-square may be defined as

$$\chi^2 = \frac{\sum\limits_{i=1}^{n} (\bar{x}_i - \bar{x})^2}{\bar{x}}$$

where x_i are the observed values and $\bar{x}$ is the sample mean. Values of chi-square are tabulated for various degrees of freedom and probability level. If the calculated χ^2 is greater than table value, our null hypothesis is rejected. Conversely, if the calculated χ^2 is less than the table value at the probability level selected, the hypothesis is accepted.

3.3 DISTRIBUTIONS WITH APPLICATION TO TRACER METHODS

Many methods have been proposed to study the distribution of experimental data. Such a distribution expresses the frequency distribution of repetitive measurements, which is all based on probability. Two such distributions frequently encountered in scientific work are the normal distribution and the Poisson distribution. The authors have previously mentioned some properties of these distributions; it is now appropriate to present a more detailed description and some applications.

3.3.1 Normal Distribution

The normal distribution is one of the most useful statistical measurements. Experimentally this distribution function has been found to adequately describe systems in which measurements under study are affected by a large number of errors all acting independently. The distribution of random fluctuations in physical phenomena conforms to this description.

The normal-distribution curve is a symmetrical bell-shaped curve which is completely determined by two parameters, μ, the mean, and σ, the standard deviation. If the sample is $n = 20$ or greater, the sample $\bar{x}$ is a good estimate of μ and s of σ. If the measurements conform to a normal distribution, one can expect that 68 per cent of the observed values lie within one standard deviation (1σ) from the mean, 95 per cent in the range $\mu \pm 2\sigma$, and 99.7 per cent in the range $\mu \pm 3\sigma$.

3.3.2 Poisson Distribution

The Poisson distribution is a discrete distribution that describes processes whose probabilities of occurrence are small yet constant; that is, some kind of event occurs repeatedly but haphazardly. Emission of radioactivity conforms to such a description. In the Poisson distribution, the variable x assumes only integer values. In contrast to the normal distribution, the Poisson distribution is defined by a single parameter, the mean μ. The standard deviation is simply

$$\sigma = \sqrt{\mu}$$

or, in terms of our sample parameters,

$$s = \sqrt{\bar{x}}$$

It is now very simple to express the error involved in a counting rate. If the counting rate were 10,000 counts/min and one counted for 1 min, the standard deviation would be $\sqrt{10,000}$ or 100 counts/min; therefore, one should express the rate as $10,000 \pm 100$ counts/min.

It is important to recognize the distinction between counts and counting rate. In the above example if one had counted for only 0.1 min at a counting rate of 10,000 counts/min, one should have obtained only 1,000 counts, and the standard deviation would have been $\sqrt{1,000}$ or 31.6 counts per each 1,000 counts. When this is expressed as a counting rate of 10,000 counts/min, the standard deviation is $10,000 \pm 316$ counts/min. Thus, the statistical significance depends on the number of counts and not on the time it took to count them;

i.e., standard deviation must be calculated on the basis of absolute counts, not counting rate.

3.4   CORRELATION AND REGRESSION

Up to now this text has been limited to the statistics involved in a single property, or single random variable. It is possible, however, to conceive of problems in which there is more than one variable. For such problems it is necessary to enter into a different phase of statistical analysis, to investigate methods that may be used to determine the interrelationship of variables. These methods are usually classified into two groups, namely, correlation and regression. Let us first consider correlation methods.

3.4.1   Correlation

Suppose two different properties of a sample group are measured. If a plot of one versus the other, or x versus y, is done, what is termed a *scatter diagram* is obtained. It may appear that there is some linear or straight-line relationship between x and y, and it is therefore useful to have a measure which expresses the extent of relationship in view of the random spread of the data (or the scatter). Such a measure is called a *correlation coefficient* and has the following form:

$$r = \frac{\Sigma (x_i - \bar{x})(y_i - \bar{y})}{n s_x s_y}$$

where $\bar{x}$ is the sample mean, x_i the observed value, s_x the standard deviation in x_i, s_y the standard deviation in y_i.

Figure 3–7 illustrates some scatter diagrams and their associated values of r. It is evident that the increasing values of r determine the increasing degree of linear relationship. If r assumes a negative value, this merely means that as x increases, y decreases. The values of r lie between -1 and $+1$, with $r = \pm 1$ only if the points lie on a straight line.

Let us now calculate r for the example to follow. Before we undertake this calculation, however, let us rearrange the formula for r into what is often called the computational form. An alternative form of r is

$$r = \frac{\Sigma (x_i - \bar{x})(y_i - \bar{y})}{\sqrt{\Sigma (x_i - \bar{x})^2 \, \Sigma (y_i - \bar{y})^2}}$$

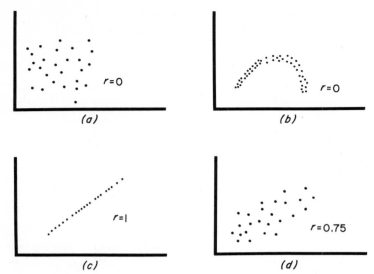

Figure 3-7.   Scatter diagrams with various *r* values.

If one performs the indicated multiplications and uses the definition of the mean,

$$r = \frac{n\Sigma xy - \Sigma x \Sigma y}{\sqrt{[n\Sigma x^2 - (\Sigma x)^2][n\Sigma y^2 - (\Sigma y)^2]}}$$

where the use of deviations is avoided, and the sums of x, y, x^2, y^2, and xy are all more readily calculable.

An example of correlation-coefficient determination is the correlation of the plasma volume by two methods in Table 3-3, which follows; x is the space determined in milliliters per kilogram obtained by using ^{131}I-RISA, and y is the space in milliliters per kilogram obtained by using [113m]In-transferrin.

$$r = \frac{20 \times 28.532 - 761 \times 714}{\sqrt{[20 \times 31,201 - (761)^2][20 \times 26,712 - (714)^2]}}$$

Hence, placing these summations in the equation for r yields $r = 0.827$. The result $r = 0.827$ shows a high degree of correlation between x and y.

The question is now posed, "Is a high degree of correlation between two variables significant?" For example, would it be significant if a high degree of correlation between incidence of lung cancer and increase in the standard of living were found? The quantity r is a pure

TABLE 3–3.
Correlation of Plasma Volume by Two Methods

x (^{131}I-RISA)	y (^{113m}In-transferrin)	x^2	y^2	xy
23	21	529	441	483
64	45	4096	2025	2880
45	40	2025	1600	1800
44	40	1936	1600	1760
51	49	2601	2401	2499
27	30	729	900	810
26	28	676	784	728
24	26	576	676	624
25	29	625	841	725
33	42	1089	1764	1386
40	43	1600	1849	1720
31	31	961	961	961
44	39	1936	1521	1716
43	41	1849	1681	1763
43	41	1849	1681	1763
35	31	1225	961	1085
29	25	841	625	725
44	29	1936	841	1276
39	38	1521	1444	1482
51	46	2601	2116	2346
$\Sigma =$ 761	714	31,201	26,712	28,532

mathematical concept and exercises a purely mathematical relationship between two variables. It is completely devoid of any cause-and-effect implications. Familiarity with the field, the system, and the experiment is paramount to successful application of correlation coefficients.

3.4.2   Regression

In the preceding section the authors have discussed a method for estimating the degree of linear relationship between two variables. More often than not, an experiment is not carried out in this manner; instead, the experimenter chooses the x values at various discrete intervals, then records the changes in y. He hopes that any relationship he finds will assist him in making predictions about y. A correlation analysis is obviously incapable of handling such predictions. The method used to handle this kind of investigation is called *regression*. In the following section will be discussed a specific method for handling the situation where x and y are linearly related.

Method of Least Squares.   Regression analysis is the fitting of a curve to a given set of experimental points. The problem is to estimate the parameters of the curve in some efficient manner. The best known of these methods is that of least squares.

The desired curve from regression analysis is used to predict or estimate values of one variable with respect to another. Therefore, the fitted curve should be such that the deviations from the curve are minimal. Absolute values are again avoided by squaring the deviations and requiring that the sum of the squares be minimized.

Let us consider the problem of fitting a straight line to a set of n points. The equation for such a straight line is

$$y_i' = a + bx_i$$

where b is the slope and a is the intercept of the line on the y axis, and where y_i represents the calculated value of the function y for the ith experimental point involving the variable x.

Now $y_i - (a + bx_i)$ equals the deviation between the measured and the calculated values of y, that is, $y_{obsd} - y_{calc}$. The sum of the squares of these differences must be a minimum. If this summation is treated like any other function, it is possible to obtain its minimum value by setting the first derivative equal to zero. For example, the derivatives with respect to a and b are

$$\frac{d\Sigma(y_i - a - bx_i)^2}{da} = \Sigma 2(y_i - a - bx_i)(-1) = 0$$

$$\frac{d\Sigma(y_i - a - bx_i)^2}{db} = \Sigma 2(y_i - a - bx_i)(-x_i) = 0$$

The results of these operations are

$$a = \frac{\Sigma x^2 \, \Sigma y - \Sigma x \, \Sigma xy}{\Sigma x^2 - (\Sigma x)^2}$$

$$b = \frac{n \, \Sigma xy - \Sigma x \, \Sigma y}{n\Sigma x^2 - (\Sigma x)^2}$$

where a and b have been rearranged into this form to simplify calculations. The fitted line would be

$$Y = \frac{(\Sigma x^2 \, \Sigma y - \Sigma x \, \Sigma xy) + (n\Sigma xy - \Sigma x \, \Sigma y)X}{n \, \Sigma x^2 - (\Sigma x)^2}$$

where Y and X represent the two variables for our curve. It should be noted that b, the coefficient of the variable, is often referred to as the *regression coefficient*. The standard deviation of b is given by

$$s_b^2 = \frac{n \, \Sigma(\Delta y)^2}{n - 2 \, [n\Sigma x^2 - (\Sigma x)^2]}$$

and of a is given by

$$s_a^2 = \frac{(\Delta y)^2 \Sigma x^2}{n - 2 [n \Sigma x^2 - (\Sigma x)^2]}$$

where $\Delta y = y_{calc} - y_{obsd}$.

The data for the plasma volume in the Table 3–3 determinations can be treated as a least-squares problem; it is found that the optimum straight line is

$$y = 12.75 + 0.60x$$

and that $s_a = 4.00$ and $s_b = 0.10$.

The variance of y values about the line is calculated in the same manner as for single variables, and in the example given it is found that $s_{yx} = 4.3$.

$$s_{yx}^2 = \frac{\Sigma (\Delta y)^2}{n - 2}$$

Figure 3–8 shows a plot of a calculated regression line with the standard deviation from regression included. If the data points are from a normal population, then it is expected that approximately

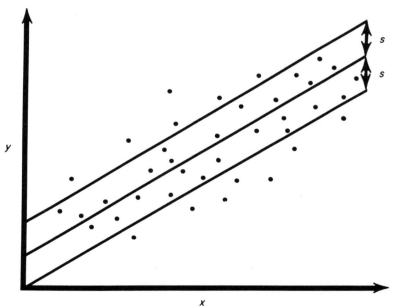

Figure 3–8. A plot of a calculated regression line with the standard deviation from regression shown as the outer two lines.

68 per cent of the experimental points will lie within the limits of one standard deviation unit from regression. It is apparent from this kind of analysis that a constant error is assumed for all points. This assumption is often not correct for certain experimental conditions. For example, if the standard deviation in Figure 3–7 is expressed as a percentage error, the percentage error is less if the points have large values. Therefore, the method of least squares favors or weights more heavily the large values in a set of points.

In many cases in which the range of x is small, the assumption of a constant error is usually satisfactory. However, in some experiments, for example, variation in concentration over several orders of magnitude, a constant error is not a good assumption. In fact, many experiments are designed in which a constant percentage error is maintained throughout the range of the experiment. Such limits are shown in Figure 3–9. There are several methods for handling this problem of weighted or favored points. One possibility is the minimization of sums of the squares of the relative deviation; that is, $\Sigma[(y - y_i)^2/y_i]$ is minimized. Another possibility is minimization of the sums of the squares of the constant percentage errors; that is, $\Sigma(y_i^2/y_i)$ is minimized. These methods are, of course, closely related. A full treatment of the weighted least-squares method may be found in most statistics books. Here it is presented only to indicate the inherent difficulty in the usual least-squares treatment.

The authors have also considered only linear regression analysis.

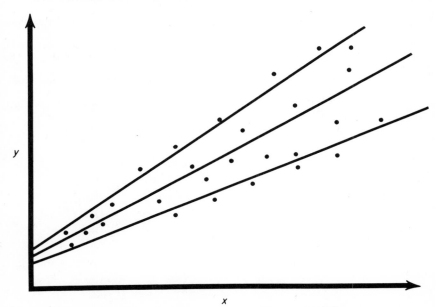

Figure 3–9. Plot of a regression line with standard deviation in a case where a constant percentage error is maintained.

It is, of course, possible to use the least-squares method on polynomials. In such a case, one would minimize the squares summation with respect to each of the coefficients of the variable raised to its appropriate power. This results in n equations in n unknowns, which may be solved simultaneously for the various coefficients. The task is a bit tedious, but with the aid of computers such calculations are now easily undertaken.

In the foregoing sections, the more common methods of statistical analysis have been presented. It must always be noted that the true mean, the true standard deviation, and the true regression line are never known, because it is generally not possible to make a measurement on the whole population. The sample mean, standard deviation, and regression line are estimates of the true mean, true standard deviation, and true regression line. By so estimating, one has also considered sample size and the probability that the sample mean is a good estimate of the true mean. It must be stressed, however, that statistical analysis is no substitute for a well-planned experiment and thorough knowledge of the system being studied.

BIBLIOGRAPHY

Alder, H. L., and E. B. Roesscer: "Introduction to Probability and Statistics," 2d ed., W. H. Freeman and Company, San Francisco, 1962.

Arkin, H., and R. R. Colton: "Tables for Statisticians," Barnes & Noble, Inc., New York, 1950.

Edwards, A. L.: "Statistical Analysis," 2d ed., Holt, Rinehart and Winston, Inc., New York, 1963.

Evans, R. D.: "The Atomic Nucleus," McGraw-Hill Book Company, New York, 1955.

Lindgren, B. W., and McElrath: "Introduction to Probability and Statistics," The Macmillan Company, New York, 1959.

Mode, E. B.: "Elements of Statistics," 3d ed., Prentice-Hall, Inc., Englewood Cliffs, N.J., 1961.

Tittle, C. W.: How to Apply Statistics to Nuclear Measurements, *Nuclear-Chicago Corp. Tech. Bull.* 14, 1964.

Chapter 4

DATA PRESENTATION

When presenting data, one is usually limited to two alternatives: (1) tabular presentation and (2) graphical representation. Tabular presentation is merely the listing of the actual data as measured and expressed in the number of significant digits appropriate to the precision of the measurement. The chief advantage of tables is their high precision. Graphical presentation, however, has alternative advantages. It displays visually the relationship between the variables. If the relationship is a smooth curve, the graph allows for the estimation of points that were not measured experimentally; in many cases it also allows for the determination of a functional equation between variables (see Sec. 3.4.2 on regression methods). Graphical analysis averages out random fluctuations in the data. The obvious disadvantage of a graph is its low precision. In order to show a general trend, one does not usually use a graphical scale that is comparable with the precision of the measured data. In other words, the graphical scale has fewer significant digits than the measured data.

Since the use of tables is probably familiar to all readers, the following discussion will stress the construction and use of graphs for data presentation.

4.1 CONSTRUCTION OF A GRAPH

There are a few simple rules for graph construction that are worthy of mention. It is customary, first of all, to plot the independent

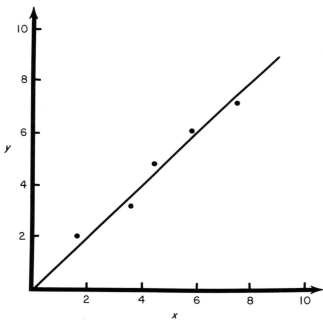

Figure 4-1. A well-constructed graph describing the experimental points shown.

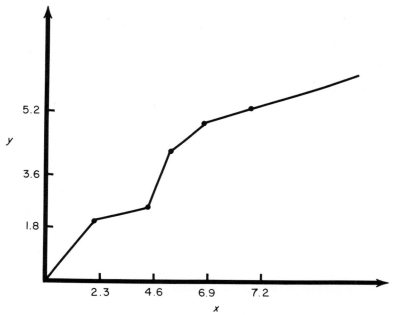

Figure 4-2. A poorly constructed graph describing the experimental points shown.

variable on the horizontal axis (abscissa), which is perpendicular to the vertical axis (ordinate). The independent variable is usually considered that variable which can be measured with the greatest precision.

Although it is not necessary for both variables to be plotted on the same scale, the choice of scales is extremely important. The range of scale not only must include the range of the experimental points but also must express sufficient precision so that any random fluctuation in the data is, at least, indicated. Graphing is an averaging process; every point need not lie on the best or most probable curve. A graph should tell the complete story; the axes must be labeled clearly and the graph must be titled. Figures 4.1 and 4.2 are examples of a well-constructed and a poorly constructed graph, respectively.

4.2   EXPERIMENTAL ERRORS

In Chapter 3 the errors involved in repeating a measurement several times were discussed. For graphical purposes, the mean of the measurements is plotted. It is customary to specify the experimental error involved in the measurement. This is done by drawing line segments of appropriate length through the mean. If the sample is sufficiently large to handle by statistical methods, the error line, or

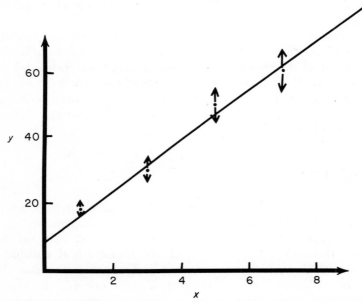

Figure 4–3. A linear plot of x versus y with standard deviation in y shown for selected values of x. In this case the variation in x is small compared with the variation in y.

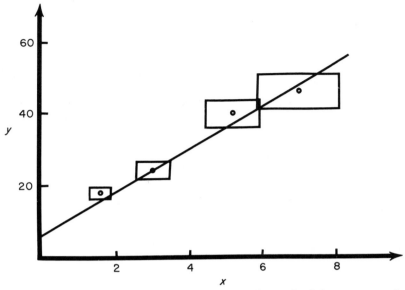

Figure 4–4. A linear plot of x versus y with standard deviations in both x and y shown. In this case the variation in both x and y is significant.

flags, are usually one standard deviation from the mean or the standard error of the mean. The error flags may, however, merely express the range of the experimental measurements.

In some cases, points on a graph are presented as the center of a rectangle whose sides represent the experimental uncertainties of the measurement of the two variables. But, as has been indicated previously, one usually selects the independent variable in such a way that its experimental uncertainty is very small compared with that of the dependent variable. Figures 4.3 and 4.4 illustrate the usual representation of experimental errors.

4.3 THE BEST CURVE

In Chapter 3 we discussed regression methods, which are used to determine the best-fitted curve for a set of experimental points. Although regression methods are usually very feasible for linear functions, the calculations involved are tedious and complex for nonlinear functions. If the plot of one variable versus a second shows a nonlinear relationship, as represented in Figure 4.5, the equation of the best curve may often be obtained by trial plots of simple powers or the reciprocal of the independent variable versus the dependent variable. Figure 4.6 demonstrates that a linear relationship was

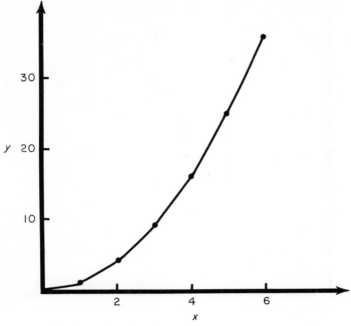

Figure 4-5. A nonlinear plot of x versus y.

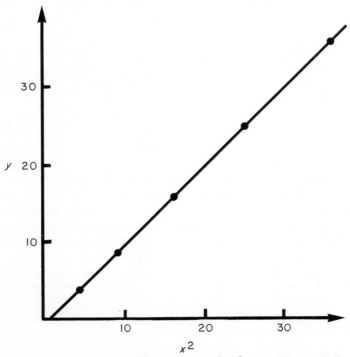

Figure 4-6. A plot of x^2 versus y for data in Figure 4-5.

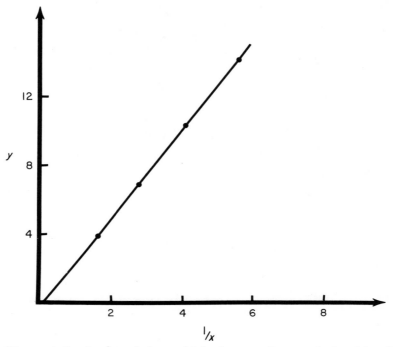

Figure 4–7. A plot of data which shows a linear relationship of $1/x$ versus y.

obtained by plotting x^2 versus y. The equation of the functional relationship between y and x is $y = ax^2$, where $a = 1$. Figure 4.7 shows a linear relationship between y and the reciprocal of x. The functional equation for Fig. 4.7 is $y = a/x$, where $a = 2.5$.

4.4   THE BEST CURVE AS AN ESTIMATOR

Regression methods are primarily geared to using the fitted curve to esimate or predict values of the dependent variable from the independent variable. There are two techniques that are used to estimate values from a curve when the actual points have not been measured. One is interpolation; if a curve is constructed for experimental points, and values of the variable which lie between the experimental points are estimated, these values have thereby been interpolated. The second technique is extrapolation, which involves extending the curve beyond the range of the experimental points. Particularly when linear relationships are involved, it is often desirable to extrapolate the curve to find the y intercept, i.e., the point where $x = 0$. This kind of extrapolation is often of great theoretical importance. An example is a chemical reaction for which product y cannot be experimentally measured in the absence of reactant x.

There are many examples in scientific endeavors in which linear extrapolation to $x = 0$ is not valid; for example, the plasma concentration of dye as a function of time, following the intravenous administration of Evans blue, has been used as a measure of plasma volume by the following formula:

$$\text{Volume of distribution} = \frac{\text{quantity administered} - \text{quantity excreted}}{\text{concentration}}$$

If samples are taken at 20 to 40 min after injection, the volume of distribution at $t = 0$ (the time of injection) cannot be evaluated by simple linear extrapolation to 0 (Fig. 4–8). By the same token, one may extrapolate the curve to values greater than those measured experimentally. Again, the actual relationship between the variables may be nonlinear beyond the range of the experimental points. In addition, one may often mistakenly conclude a linear relationship between variables if the range is too small; i.e., a small portion of curve may appear linear though it is, in fact, nonlinear. To avoid such pitfalls, there is no substitute for well-planned experiments over reasonable ranges and familiarity with the system under investigation.

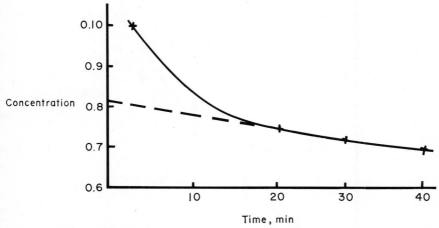

Figure 4–8. Plasma concentration of Evans blue dye as a function of time, following intravenous injection. Note that if samples were taken between 20 and 40 min, a linear extrapolation to 0 time would not represent the true dye concentration at 0 time since the curve is not linear over the first 20 min. Actually there is a much greater initial concentration than would be apparent from a linear extrapolation of the terminal slope.

4.5 THE MANIA FOR SEMILOG PLOTS

In tracer studies special attention should be paid to two particular types of graphs, the semilog and the log-log plot. Many data from the use of tracer experiments result in the fitted curves of power functions, for example, radioactive decay and rates of chemical reactions. It is often very convenient to plot the logarithm of the dependent or independent variable, or both, depending, of course, upon the particular function. Logarithmic paper provides a fast, easy method for such plots. The lines ruled off on logarithmic paper are proportional to the logarithms between the numbers 1 and 10. If only one axis is logarithmic, whereas the other is linear, the graph paper is called semilog paper. If both axes are logarithmic, the graph paper is called log-log paper. Logarithmic graph paper enables one to plot logarithms without use of logarithm tables.

It has been indicated previously that the use of graphical analysis of data sacrifices precision compared with a tabular presentation. The use of log-log or semilog paper reduces precision even lower than that which can be obtained with ordinary linear graph paper.

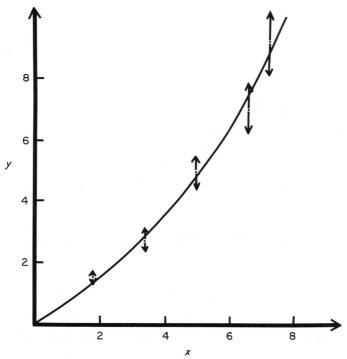

Figure 4–9. A plot of some experimental data on linear graph paper.

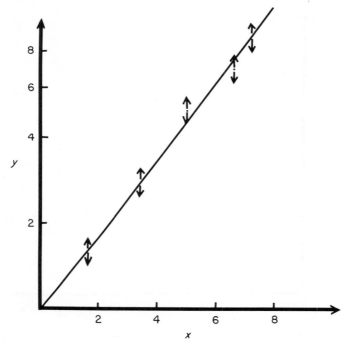

Figure 4–10. A plot of the data shown in Figure 4–9 on semilog paper.

Although logarithms can be expressed very accurately to at least four significant digits, their graphical representation on logarithmic paper is, at best, only a fair approximation. Logarithmic plots further smear out differences between numbers and visual representation of the experimental spread of the data. Log-log or semilog plots often appear to give a straight line. Figures 4.9 and 4.10 illustrate the differences between plotting a particular set of data on logarithmic paper and on ordinary graph paper. It is interesting to note that Figure 4.9 is a randomly constructed curve which yielded a straight line when plotted on semilog paper. In addition, error flags have little significance on log-log and semilog plots.

If a straight line is obtained on a log-log plot, the equation for the functional relationship between the variables is given by $y = ax^n$. It can be seen that this equation may be rewritten in terms of logarithms,

$$\log y = \log a + n \log x$$

where n is the slope of the line, and $\log a$ is the y-axis intercept. The slope n is found by actually measuring the geometrical slope, as shown in Figure 4.11 ($n = h/r$).

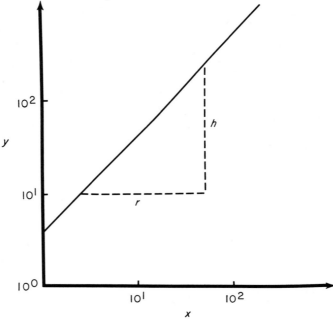

Figure 4–11. Graphical method of determining the slope of a line plotted on log-log paper. Slope is given by $n = h/r$.

Semilog plots are particularly useful for equations of the form $y = ae^{nx}$. This is, of course, the form of radioactive-decay and some reaction-rate equations. One can easily express the above equation to the base 10; that is,

$$2.3 \log y = 2.3 \log a + nx$$

and plot it on semilog paper. Again, if our experimental data give a straight line on a semilog plot, n is the slope of the line. Confusion often arises about how to determine the slope of a semilogarithmic plot. Two points must be considered: (1) the slope is the change in y per change in x and so must be expressed in units of x; (2) the apparent geometric slope must be corrected by the grid scale if the grid is a rectangle instead of a square. This is shown in Figure 4.12. In many cases, it is simpler to calculate the slope by merely noting the change in y for a particular change in x and looking up the actual logarithm. For example, in Figure 4.12 for Δx equal to $25 - 6 = 19$, y at $x = 25$ is 400; at $x = 6$, $y = 10$. Therefore, the slope equals

$$\frac{\log 400 - \log 10}{25 - 6} = \frac{2.301 - 1}{19} = \frac{1.301}{19} = 0.0697$$

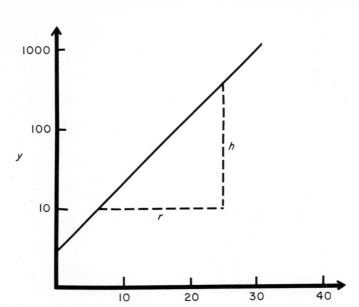

Figure 4–12. Graphical method of determining the slope of a line plotted on semilog paper. Slope is given by $n = h/r$.

Again the authors cannot emphasize enough that although many kinds of mathematical analysis of an experimental system are powerful tools, they are not a substitute for scientific acumen and familiarity with the system under investigation. Mathematical relationship does not express any cause-or-effect basis, but merely a logical system of analysis. The experimenter's scientific knowledge and prowess are essential to meaningful conclusions from data collected.

BIBLIOGRAPHY

Kruglak, H., and J. T. Moore: "Basic Mathematics for the Physical Sciences," McGraw-Hill Book Company, New York, 1963.

Worthing, A. G., and J. Geffner: "Treatment of Experimental Data," John Wiley & Sons, Inc., New York, 1943.

Chapter 5

EXPERIMENTAL DESIGN

5.1   SCIENTIFIC METHOD

In the preceding chapters tools have been presented that can be used to make observations more explicit by means of suitable definitions and mathematical propositions. The formal character and relationships of the observations can be developed deductively in order to predict certain consequences. To formalize these observations is an intricate part of what is regarded as science; however, at some point, one must connect mathematical concepts to reality. Science, in the last analysis, depends upon observation or experience in the external world.

We cannot overemphasize the importance of an elegant, precisely-designed experiment in tracer studies or any other area of science. The biological and medical sciences are still largely observational and experimental. They do not have the extensive theoretical frameworks of the physical sciences. In the biomedical sciences well-designed experiments and the interpretation of the observations are more difficult than in the physical sciences because of the greater complexity of the elements involved. *Observations using the tracer method are particularly prone to erroneous conclusions, primarily because of excessive involvement with the actual process of data gathering and too little consideration for experimental design.* It is to this end that we shall discuss not only experimental design, but also that whole process by which one acquires knowledge about the physical world—the scientific method.

94

Let us, however, first consider a definition of the scientific method so that the discussion may proceed from some foundation. A dictionary definition is as good as any other, for one is faced with defining a dynamic process in static terms. The problem is analogous to some modern art forms, which in order to define one must experience. The active experience is part of the definition. The Random House Dictionary states the following about the scientific method: "A method of research in which a problem is identified, relevant data are gathered, a hypothesis is formulated from these data and the hypothesis is empirically tested." Although the research conditions are admittedly quite different for the mathematician, the theoretician, and the experimentalist, the scientific method appears to be the same in all sciences.

Despite significant recent advances in the development of biomedical sciences, these disciplines are still, as stated previously, generally observational and experimental. One must recognize that the near perfection of observational and measuring devices has expanded the realm of knowledge in biomedical sciences so that in some particular areas research conditions approach the physical sciences. However, there are still vast areas of study in biology and medicine in which phenomena can only be observed qualitatively, and any theoretical framework is mere classification of observations. The following discussion will be confined to the research conditions that are peculiar to the observational and experimental sciences.

Our dictionary definition has divided the scientific method into four realms of activity. Each "step" in the method presupposes that the "user" has inherent scientific qualities, such as powers of observation and intuition, and some creativity. Each step presents different kinds of difficulties.

The whole process begins with the identification of a problem to be investigated. The scientist may either notice gaps in the present body of knowledge, which may encompass whole areas of study (for example, the composition of genetic material or subatomic particles) or he is faced with inconsistencies when a fact or a series of facts apparently do not agree with previously established theories or accepted explanations. These facts may arise from data collected from a systematic observation of a phenomenon (common in biology and medicine) or from an accidental finding (an unexpected result from some other-directed observations), or perhaps from the reexamination of some previous results in the light of the consequences of a new theory. One of the most common faults in research is the failure to identify the problem; we shall further discuss this fault, which is termed phenomenology, in a subsequent section of this chapter.

Although no strict rules can be formulated for identifying a problem suitable for investigation, one can, perhaps, be guided by

Descartes in his "Discourse on Method":

> . . . the first was never to accept anything for true
> which I did not clearly know to be such; that
> is to say, carefully to avoid precipitancy and
> prejudice. . . .

The essential point is the necessity of systematic doubt and criticism, of complete mental freedom even to confront generally held theories or concepts. The successful scientist must never become so submerged in the theories and practices of the day that he can no longer critically assess or reevaluate them. Indeed, the cast of doubt may open new avenues of discovery.

Once the problem has been stated, the next step is to ascertain what investigations have already been done — on the specific problem or on related problems. It is at this point that the experimentalist must engage in a thorough survey of the scientific literature. Recent review articles are often helpful and not only provide a summary of the present "state of the art" but also a list of primary references from which to proceed. There is, however, no substitute for consulting original articles for the exact data as presented by the investigators. With the growth of great numbers of scientific journals and the articles contained therein, the task of a complete literature search has become increasingly more difficult. Computerized, key-title-word indexes are helpful for recent work; abstract indexes and abstracts, for earlier work. The experimenter must realize the importance of a thorough literature search for the gathering of all existing pertinent data prior to actually undertaking the experiment. Too often he finds that much time and equipment could have been saved if he had taken the time to gather all the existing data. It should also be noted that a researcher should keep abreast of the literature in a wide area of related fields to ascertain any new techniques or principles that may be applicable to his own work.

Once the pertinent data have been collected, one can form a hypothesis. A hypothesis is an explanation, a generalization of the "facts," from which one may predict certain consequences to be tested. Hypotheses may take the form either of mathematical equations which can be interpreted to fit the facts or of simply ordering and classifying experimental data. The former is preferred, for it presupposes that one has assigned a numerical scale to various phenomena, and the mathematical process of deduction allows the prediction of interesting consequences from the hypothesis. The second form of the hypothesis requires a newer mathematics which has been developed only in recent years.

There are two more difficult problems with which one has to deal in hypothesis making. One is the actual formation of a possible hypothesis. Quite clearly, there can be infinite numbers of hypotheses which will fit certain facts. One obvious problem is how hypotheses are formed in the first place. This philosophical question is not within the scope of this book, for there is no simple answer. Hypothesis making is a creative process, a product of the scientist's creative genius.

The second remaining problem concerned with hypothesis-making is how to choose among several hypotheses; toward which one do we direct the actual experiments? It is generally accepted that the simplest one is provisionally chosen, "on the supposition that this is the more likely to lead in the direction of truth" (Descartes). From the hypothesis there follow certain consequences (predictions); one must then test and verify our predictions, i.e., return to basic observations to see whether the hypothesis is "proved." It must be noted that hypotheses are never really proved but are assigned degrees of credibility or probability based on verification of a prediction. Observations and predictions are only approximate (experimental error); therefore, they are actually probability statements. An unfavorable result can only make a hypothesis unlikely or improbable, but not certainly false.

Each test, then, modifies the credibility of a hypothesis. If the credibility of the chosen hypothesis becomes less than that of a competitive hypothesis, it is rejected in favor of the alternative. Later experiments may also eliminate the alternative; i.e., the process does not stop, but the facts (observations) ending one cycle of the scientific procedure form the beginning of the next. It is possible that difficulties will arise when trying to indicate that one hypothesis is simpler than another. If a number of alternative hypotheses are expressed, each appearing to be about the same probability on the basis of the data, each having certain consequences (predictions), it is necessary to perform experiments that test the credibility of those alternatives. Reject those whose credibility is lower than others. The rule of thumb, however, is still that the simplest hypothesis is given the highest credibility.

At this point in a discussion of the scientific method, one must design and specify experiments to test the predictions of the hypothesis. Later sections of this chapter will consider specific types of experimental procedures. Let us here, however, discuss some general ideas and guidelines on observations and experimentation.

Active observation is always a kind of experiment, but it also involves selection of "pertinent facts" from the whole of experience. An experiment is a controlled observation in which the variables are finite and known. The scientist assumes that in considering a small portion of the universe, he has an isolated system in which he can neglect all the rest of the universe. The ideal is, of course, never attained, but it is approached by altering the field of observation (views) until the isolated system behaves as though it were really isolated. The traditional method is to vary only one factor at a time while all other variables are held constant. Biological systems, in fact, present a complicated situation for this kind of approach. Indeed, it has been suggested that this method of isolation of a single factor often occurs under restricted and unduly simplified conditions. Recently, there has been emphasis on planning experiments to test a number of variables at the same time. Such experiments examine each variable in the light of a variety of circumstances, and interactions between variables may be detected. The appropriate mathematical and statistical techniques enable more than one variable to be included in one experiment. Surely both approaches, single and multi-variable analyses, are valuable to appropriate problems.

Before a large-scale experiment is planned, it is often profitable to conduct a small-scale preliminary experiment that can give provisional indications as to possible verification of the hypothesis. Often, such a preliminary test is done under extreme conditions; for example, in a tracer experiment by using extremely large amounts of the tracer to see if there is a good chance of successful results.

Another point that cannot be overemphasized and that will be discussed in more detail later is that the experimenter must have a full understanding of his technical methods and instruments and of their limitations and degree of accuracy.

Finally, the essence of a good experiment is that it should be reproducible. The inherent variability of biological materials often challenges reproducibility. However, as suggested in a previous chapter, the powerful techniques of statistics have helped the experimenter to meet the challenge, providing an estimation of variability and margin of error that can be tolerated.

We have not, at this point, distinguished precisely between many terms that swell the jargon of the scientific method, e.g., hypothesis, theory, law, fact, premise, and postulate, all of which are generalizations. In many cases, such terms are used interchangeably, and indeed the distinction is sometimes artificial. It is, however, helpful to attempt definitions of some of the more familiar yet often evasive terms. Some useful distinctions are based on degrees of probability or are in relation to remoteness or nearness to experience (observation).

A *fact* is usually supposed to mean a particular event (observation), which is, in a sense, experience. It is the first and simplest generalization of experience. (Unfortunately, the word is most often used ambiguously; i.e., "facts" are usually statements about the external world to which a great degree of confidence has been attached, whether particular or general. In this latter case, the assertion is more properly a *law*, rather than a report of the occurrence of an event.)

A *law* is a statement of a relationship in which each member must be independently measurable. A law is a combination of facts in a wider generalization. Laws are our abbreviations of experience and are generalizations of high probability. A *hypothesis* may be considered a law of low probability; it lacks the verification step.

A *theory* is further removed from direct experience. It may not contain any observational terms at all. Theories are more general than laws and, indeed, may incorporate many laws.

All these concepts combine to present a hierarchy of explanation, in which facts on the lowest level are explained by laws and theories; each theory in turn is explained by the theories on a higher level until the limits of our present knowledge are reached.

5.2   TYPES OF EXPERIMENTS

5.2.1   Phenomenology

In the previous section the authors attempted to outline the scientific method and to suggest at least its systematic rigors and the importance of a well-constructed, delicately detailed experiment to "verify" the proposed theory. It is appropriate to turn now to various experimental approaches. Let us first examine an experimental method that is particularly common in biology and medicine, and indeed, in areas of the physical sciences that are still primarily observational. It is called *phenomenology*. Basically, phenomenology is empiricism with a greater or lesser degree of actual conviction to that specific school of thought. Empiricism maintains that all knowledge must originate from experience, which, although encompassing a whole philosophy, in practical terms means that our frontiers of knowledge are extended only through actual observation and experience, by a trial-and-error procedure. Perhaps the empiricist sounds synonymous to the experimenter; the crucial point here is that they are not mutually exclusive. The "experimenter" as we have described him is performing purposeful, controlled, selective operations in which he is an observer; the "empiricist," in our sense, is merely an observer in a situation where a trial-and-error technique is employed in a loosely purposed pseudo-experiment. This kind of definition will surely offend those who are committed to empiricism as a school of philosophical thought, but it is given to point out a kind of research method that is very common to biology and medicine, and to detail some of its pitfalls.

By phenomenology the authors mean certain types of experiments by which "we see what we can see," i.e., an observation for the sake of observation. "Let's try this drug; it can't do any harm" is the kind of reasoning process that, in general, does not contribute to the progress of medicine. The biological sciences lend themselves easily to phenomenological experiments. Present technology is such that many phenomena which were heretofore unobservable are now discernible. There are many opportunities to "see what we can see" because previously no one could ever observe the structures or events. The availability of new instruments stimulates many phenomenological experiments. The use of radioactive tracers is particularly prone to such experimental techniques, especially with the advent of highly sophisticated instruments and new isotopes. Such approaches, unfortunately, often become mere data gathering and the experimenter, a technician, simply a data collector. Phenomenology tends to generate random numbers. A well-designed experiment must have controlled or specified variables; often the numbers obtained from the "new instruments" are random because the variables were either uncontrolled or unknown. Users are often unaware of not only the limitations of the instrument but also what they have measured or observed.

It is easy to see that data analysis assumes great importance with the phenomenological approach. The emergence of such disciplines as biostatistics and systems analysis gives evidence to this fact, although it has already been pointed out that statistics can be a powerful tool in biological research.

Another aspect of phenomenology is what may be termed *serendipity*. A number of examples of accidental observations leading to fundamental discoveries has given credence to the opinion that chance plays a primary role in scientific discovery; for example, the celebrated Newton's apple. Let us examine two less legendary occasions of serendipity and its role in experimentation.

1. In 1895 Roentgen discovered x-rays. While studying cathode rays from an evacuated tube covered with black cardboard, he noticed fluorescence on a screen of barium platinocyanide that lay on the table near the tube. He pursued the matter further and found that the mysterious rays which had produced the fluorescence came from his cathode tube. They had passed through the paper, which was opaque to all other forms of light known at that time. A systematic study ensued during which he investigated the properties of this new penetrating light. He found that it affected photographic plates and also found accidentally that the new rays would outline the bone structure in his hand.

2. Sir Alexander Fleming was studying mutations in some colonies of staphylococci in 1928 and noticed one day that the plates

of one of his cultures had been contaminated by a microorganism from the outside air. He pursued further and noted that the colonies of staphylococci that had been attacked by the microscopic fungi had become transparent in a large region around the initial zone of contamination. Fleming reasoned that this effect could only be due to an antibacterial substance secreted by the foreign microorganism. In the months that followed he studied the principal properties of the antibacterial substance and published his observations describing penicillin in 1929.

The notable point in both these examples is not that the circumstances were accidental but that they were not exceptional. Investigators other than Roentgen had already noticed similar phenomena without attaching any importance to them. How frequent is the mishap that Fleming appreciated in its full importance. Serendipitous discoveries are, in a sense, chance happenings; but more properly, the chance occurs to that research worker who is ready to accept the significance of the phenomenon. Science, however, cannot progress by awaiting serendipitous occurrences but only by systematic pursuit during which the alert mind of the researcher discerns the unexpected observations as technical difficulties or truly unforeseen discoveries.

Again, we return to the previous conclusion that purposeful, selective experimentation in answer to specific questions or hypotheses produces the most fruitful path to successful research.

5.2.2   Problem Solving

Another method of research is what may be termed *problem solving.* As will be seen, the problem-solving method is a variation on the scientific-method theme.

The problem-solving approach may be outlined as follows: Identify the problem, pose a question, list possible answers, and design the experiment that will verify the most probable answer(s). The method consists of a systematic dissection of a large problem, breaking it into components that can be evaluated, then synthesizing the general answer from the experimental result. Descartes expresses this approach in his "Discourse on Method" with his third law: "... conduct my thoughts in such order that by communing with objects the simplest and easiest to know, I might ascend little by little, and as it were, step by step, to the knowledge of the more complex; assigning it through a certain order even to those objects which in their own nature do not stand in relation of antecedence and sequence."

Much time and emphasis are given to designing a specific experiment to demonstrate the answer or answers to a specific question.

One should not be concerned with generating data to "see what we can see" during which the experiments are unselective and uncontrolled. There are many so-called laws of nature that are asserted with the utmost confidence and are based upon a minute number of observations and measurements. The verification of a hypothesis or the answer to a question depends not upon the numbers of observations but upon the way in which the measurement is carried out. One critical, selectively-designed experiment can equal in value a volume of randomly generated data. There are problems, however, where the number of measurements is of great importance, that is to say, the only way in which one can express confidence in the results is to repeat the measurement many times so that negative results have not escaped. The well-designed experiment minimizes the chance of error and of overlooking the negative instance.

In general, the problem-solving approach has a slightly different emphasis compared to the "scientific method," as previously outlined. Problem solving is particularly successful in research laboratories populated with teams of scientists from different fields and trained in different techniques and points of view. Its success should be well understood, for today one is seeing discoveries that are the collective work of many scientists. In general, there is an increasingly collective character to modern scientific research.

How to Pose a Question? There are a few simple rules to keep in mind when posing a question. First and foremost, identify the problem—know what you want to find out, what you are seeking. Second, clearly, concisely, and critically develop the best question that can be addressed to the problem. Many questions can be posed for a particular problem, but the right question can open many new paths and approaches. Thirdly, list the possible answers and do not begin experiments until as many answers as possible to the question have been thoroughly considered.

How to Seek an Answer? After thoroughly considering the possible answers to the question, "What experiment gives the strongest support for the most probable and/or most answers?", the answers themselves will suggest the measurements that must be made. The researcher must acquaint himself with the various methods available for a particular measurement. Be acquainted with the technical level of instruments, with what the instruments actually measure, and with how their measurement correlates with the particular phenomenon of interest. Although we have, perhaps, suggested an obsession with the use of new instruments to study known processes, we must at the same time stress the essential role that instruments play in experimental discoveries. Some of the best-designed experiments have not yielded significant observations because the state of the instrumentation was too rudimentary to produce the expected results. Each improvement upon an instrument *may* lead to corresponding new discoveries; and of course, new instruments open paths to previously unexplored realms. On the other hand, no instrument can substitute for a profound understanding of all aspects of the phenomenon of interest.

It is the scientist's great ingenuity and keen awareness of physical reality that must guide him in designing the precise experiment to verify the most probable answer(s). The researcher must possess keen gifts of intuition and observation so that he can discern whether any unforeseen event is due to experimental difficulties or to a new fact that may well lead to a new discovery.

Once the measurement or observation has been made, one must draw conclusions from the experimental data. Care must be taken to assign cause from the experimental observations. If statistical analysis is necessary, use the appropriate controls and tests. Do not place much confidence in negative results in small-sized samples. Remember that statistics is a mathematical tool and contains no cause-and-effect consequences.

5.3 EXPERIMENTAL DESIGNS

The material-and-method section of most biological papers only skims the surface of a vast mass of practical knowledge which is transmitted between practitioners mainly by apprenticeship, demonstration, and conversation. It can be troublesome for the outsider to break into such a predominantly oral tradition.

At a time when the opportunities for excellent research in biology and medicine have never been brighter, there can be no substitution for the rigors of the scientific method which have successfully guided the physical sciences to excellence over several decades.

5.4 AN EXAMPLE

We have discussed the use of the scientific method as our guide for successful and meaningful experimentation. So far, the discussion has been mainly in the form of theoretical statements. Let us now consider a specific example of a classic work in biology done by Hershey and Chase in 1952. This is selected as an elegant tracer experiment in basic biology in which the use of radioisotopes is emphasized. Figure 5–1 is a flow diagram that illustrates the various steps and developments of this experiment. Using the problem-solving technique, let us proceed step by step to illustrate the cyclic nature of the scientific method.

Figure 5–1

GENETIC MATERIAL (or GENOME)

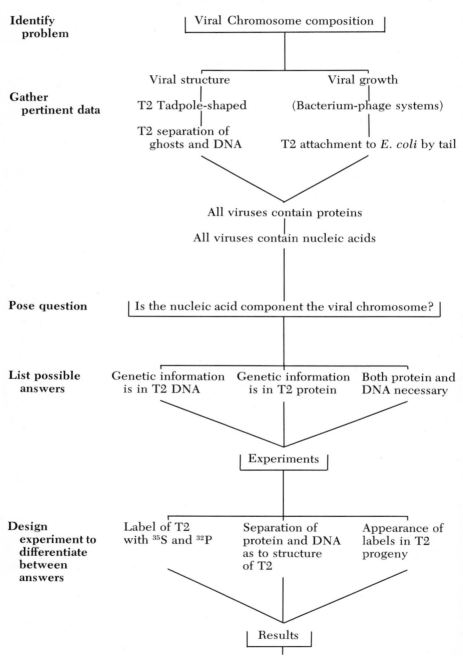

Figure 5–1 (Continued)

Plasmolysis separates resting T2 into two parts:
one containing nearly all phage sulfur and one consisting
of a solution of phage DNA

Ghosts adsorb to bacteria; DNA does not

**Interpretation
of results**

Agitated suspensions of infected cells release 75%
phage sulfur, 15% phage phosphorus. Cells remain
capable of yielding phage progeny

Phage progeny contained <1% parental sulfur
and >30% parental phosphorus

| Conclusions |

Ghosts represent protein coats that surround
DNA of intact phage particles and carry organ
for attachment to bacteria

When T2 attaches to bacterium, most of phage
DNA enters cell; residue, containing at least
75% sulfur-containing protein of phage, remains
at cell surface

Protein, membrane plays no further role in
infection after attachment to bacterium; protein
probably has no function in the growth of
intracellular phage

T2 can be separated into genetic and nongenetic
components

| Unanswered questions |

What is function of remaining 20% sulfur-containing
protein?

Does any sulfur-free phage material other than
DNA enter cell?

If so, is it transferred to phage progeny?

Is transference of phosphorus (or any other material)
direct or indirect?

First of all, identify the problem. The problem in this case is, "What is the composition of (viral) genetic material?" Obviously, this is a very general problem, and conceivably there could be many approaches that might give a satisfactory answer. The bacterium-phage system presents an excellent way to study viral growth. Viruses are parasites on the genetic level. Bacteriophage (phage) infect bacteria and multiply (replicate) themselves. Of the various known viruses the bacteriophage T2 had already been the subject of much experimentation before this work by Hershey and Chase.

It was known that T2 is tadpole-shaped and that it could be separated into a DNA fraction and a non-DNA fraction (ghosts). It was also known at the time that T2 attached itself to the bacterium *Escherichia coli (E. coli)* by its tail. The assembled information supported two statements: (1) that all viruses contain proteins; and (2) that all viruses contain nucleic acids. The following question for the particular system could be posed: "Is the nucleic acid component the viral genetic material?" There are at least three possible answers to the question: (1) The genetic information is in the T2 DNA; (2) the genetic information is in the T2 protein; and (3) the genetic information is in both the T2 DNA and the protein. At this point could an experiment or series of experiments be designed that would support one of the possible answers? It is here in the design process that the experimenter must look for the best and most appropriate tools at his disposal. Radiotracers were the precise tool that Hershey and Chase used for their experiments.

The T2 phage was labeled with ^{35}S (sulfur-containing protein) and ^{32}P (phosphate containing DNA). The subsequent experiments then separated the protein and the DNA with respect to the actual structure of T2. Where in the T2 structure was the protein? Where was the DNA? Experiments were performed to determine appearance of labels in the progeny. The actual experiment included separation of the ghost and DNA fractions of T2 followed by a specific analysis and identification of each fraction, with the same detailed analysis on the progeny. The following results were obtained:

1. Osmotic shock separates resting T2 into two parts, one containing nearly all the phage sulfur and the other consisting of a solution of phage DNA.

2. The ghost phage adsorbs to the bacteria; phage DNA does not adsorb.

3. Agitated suspensions of infected cells release 75 per cent phage sulfur and 15 per cent phosphorus; yet the cells remain capable of yielding phage progeny.

4. Phage progeny contain less than 10 per cent parental sulfur and more than 30 per cent parental phosphorus.

Now we are ready to draw some general conclusions from the

experimental results. The ghosts represent the coats that surround the T2 and are protein in composition. This protein coat surrounds the DNA of the intact phage particles. The protein coat carries the organ for attachment of T2 to the bacterium. When T2 attaches to the bacterium, most of the phage DNA enters the cell, whereas the coat, containing 70 to 80 per cent sulfur-containing protein of the phage, remains at the cell surface. This protein plays no further role in infection after attachment to the bacterium. Therefore, protein probably has no function in the growth of the intracellular phage. T2 can thus be separated into genetic and nongenetic components.

Considering the example at hand, we have now reached the stage in the scientific method where its cyclic nature is apparent. The data gathered become the "facts" for further hypothesis and experimentation. The experimental results have lent most credibility or probability to the "answer" or "hypothesis" that the genetic information is contained in the T2 DNA but have not eliminated the possibility that some protein may play a role, namely, the 20 per cent sulfur not obtained from the protein coats. The experiments have, in fact, assigned probability to more than one "hypothesis." One can therefore already pose unanswered questions that must be answered through further experimentation; that is, the whole process begins again. Some unanswered questions are as follows: (1) What is the function of the remaining 20 per cent sulfur-containing protein? (2) Does any sulfur-free phage material other than DNA enter the cell? (3) If so, is it transferred to phage progeny? (4) Is transference of phosphorus (or any other material) direct or indirect? Thus ended one phase of the chain of events that has led to the present state of knowledge of molecular biology.

We have attempted to illustrate how the scientific method is an unending process and how systematic analysis and intimate understanding of the problem and the experimental possibilities, both technical and theoretical, can facilitate the march on the frontiers of knowledge. These concepts cannot be overemphasized, especially in an age when scientific data are produced in such volumes that information communication has reached critical proportions.

BIBLIOGRAPHY

Campbell, N.: "What Is Science?" Dover Publications, Inc., New York, 1952.

Commins, S., and R. N. Linscott, (eds.): "The Philosophers of Science," Random House, Inc., New York, 1947.

Hartnett, W. E.: "An Introduction to the Fundamental Concepts of Analysis," John Wiley & Sons, Inc., New York, 1964.

Hershey, A. D., and M. Chase: Independent Functions of Viral Protein and Nucleic Acid in Growth of Bacteriophage, *J. Gen. Physiol.*, **36**:39 (1952).

Chapter 6

QUALITATIVE BIOMEDICAL EXAMPLES OF TRACER KINETICS

The preceding five chapters have dealt with tracers, the design of tracer experiments, and the mathematical tools needed to interpret the data collected. It is now time to focus attention upon the dynamics of tracer studies and their interpretation, that is, the information that the observation of a *time-variant (kinetic) process* can give us about the biomedical system under study. All tracer studies are time-dependent to varying degrees. It is just this time dependency that makes tracer studies so valuable in analyzing biomedical processes. Life is a continuously changing phenomenon, and the capacity to observe time-dependent processes by nondestructive means (e.g., tracer techniques) provides functional information which cannot otherwise be obtained. Kinetic studies, including tracer kinetics, are important in many fields of science and are principally used to answer two questions: (1) What is the rate at which a particular reaction or process occurs? (2) By what path or mechanism does the reaction occur? Investigating the latter question is often referred to as model building.

The extreme complexity of biological systems tends to confuse the treatment of tracer kinetics. This chapter will contain several qualitative examples of tracer kinetics, which will be more rigorously developed in Chapters 7 and 8.

6.1 A BIOMEDICAL EXAMPLE—THE MEASUREMENT OF BLOOD FLOW

We begin this discussion of kinetics by analyzing the measurement of blood flow. A thorough understanding of the problems involved in assessing tracer transport is prerequisite to understanding tracer-method applications in the living animal. The concepts involved here are basic to many tracer studies and will be discussed in a qualitative manner.

Historically, many methods have been advocated to measure blood flow. Early studies entailed cutting through a blood vessel and measuring the flow of blood into a receptacle, usually a graduated cylinder. Although this method allowed for an accurate determination of the amount of blood flowing into the receptacle, it unfortunately also resulted in exsanguination of the experimental animal, a process which undoubtedly altered the blood flow. Flowmeters of various types have been used in experimental situations. Although they have provided considerable physiological information, they require direct manipulation of the vessel, which again may alter blood flow.

6.1.1 The Fick Principle

Most tracer methods used to measure blood flow are based on the Fick principle, which was conceived by Adolf Fick in 1870 according to the following assumptions:

1. Any blood sample must be representative of blood flowing into and out of the vascular system being measured.

2. The total amount of indicator substance that is added to or removed from the system during a finite time as it passes through a vascular bed must be measurable.

The Fick principle, which is really a restatement of the law of conservation of mass, states that flow equals the amount of substance added per unit time divided by the concentration difference produced by this addition $(F = \dfrac{Q_t}{[C_A] - [C_V]}$, where Q_t equals amount of substance per unit time, $[C_A]$ is the arterial concentration, and $[C_V]$ is the venous concentration). This principle was first applied to assess lung blood flow by using oxygen uptake as a measure of cardiac output. Other tracers are equally applicable.

Flow methods are based on the Stewart-Hamilton application of the Fick principle. Such methods include:

1. Indicator dilution method, i.e., nondiffusible-indicator approach

2. Clearance method

3. Inert diffusible indicators.

6.1.2   Indicator Dilution Methods

Intuitively, one might think the injection of a marker (tracer) into a flowing stream and the determination of its appearance at a point beyond (i.e., distal to) the point of injection would result in a measurement of blood flow. This simple, intuitive approach poses some interesting problems. First, it is important that a rapid injection result in a "spike" (i.e., infinitely fast in relation to the flow) input into the system, since anything other than this will distort the ultimate appearance of the indicator substance at the downstream point. In biological systems this square wave or spike input is frequently difficult to attain, and therefore one speaks in terms of input "function." Such functions are mathematical descriptions of the input curve. Ideally an input should be a single straight line, and this line could then be observed at a point distal to the input. However, even with this ideal ("spike") input in a simple system (a single or branched tube), a tracer spike is not observed distal to the injection site, because the tracer must be distributed through a volume in order to reach the distal point (Fig. 6–1). Thus, the observed output function will be dependent upon the input function and the volume of the system, as well as the flow through the system. Obviously, the result is a complex situation, and usually it is not possible to dissect the various components from

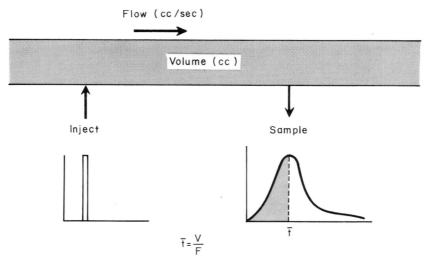

Figure 6–1. When injecting a tracer into any flowing system and sampling downstream, the change in shape of the curve—from a spike (square wave) input to a broad output—depends upon both the volume through which the tracer distributes and the flow through the system. The mean transit time from the curve (when half of the tracer reached the sampling point) is the most useful parameter to describe the system.

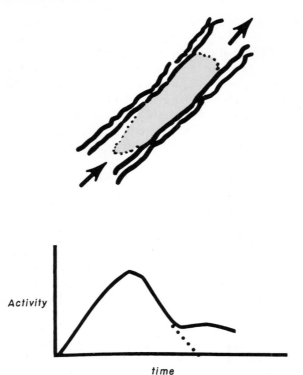

Activity

time

Figure 6-2. Curve obtained by observing the passage of a nondiffusible indicator through a vessel. The activity rises, then falls. Usually there is some second hump or early plateau due to recirculation. To measure blood flow from this technique requires knowledge of the area under the curve and extrapolation of the negative slope to zero. By external techniques, e.g., with radioactive tracers, it is extremely difficult to quantitate flow by this method since one usually does not know the volume of isotope distribution. This curve gives information of mean transit time, and mean transit time is related to flow by the equation:

$$\text{Mean transit time} = \frac{\text{volume}}{\text{flow}}$$

one another. In summation, one cannot obtain a meaningful measurement of flow by only injecting a marker, despite the intuitive observation that it should be simple.

The concepts of input function, volume of tracer distribution, and flow are extremely important to many biological processes applying tracer kinetics, as will be obvious throughout the remainder of this book.

Nondiffusible-indicator techniques (Fig. 6-2) (e.g., cardiogreen to measure cardiac output, the transit of contrast material in arteriography, and rapid sequential scintiphotography of a radioactive bolus) are all based on the simple aforementioned scheme and therefore require knowledge of input function and volume if flow is to be determined. When cardiogreen is used to measure the cardiac output, the "quantity" of color can be accurately determined in terms of concentration of the dye per unit volume of blood. This indicator is rapidly cleared (removed from the blood) by the liver, thereby decreasing the problem of recirculation. If cardiogreen is rapidly injected into the heart, and subsequently the volume and distribution assayed in an artery, the flow through this system (i.e., cardiac output) can be reliably measured.

An example of the indicator dilution approach to measure blood flow is seen in Figure 6–3, where the first peak represents the arterial dye concentration in the first pass through the system, and the subsequent peak is due to recirculation. Since it can be assumed that at the time of the first transit there is no recirculation contributing to the curve, the descending slope of the curve is extrapolated to zero. The mean arterial concentration can be calculated by integrating the curve of indicator concentration versus time (alternate: by finding the area under the curve of . . .) and then dividing by the amount of time

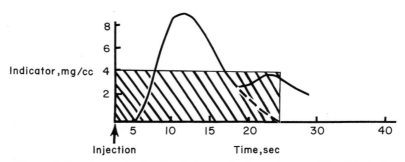

Figure 6–3. A standard calculation technique for nondiffusible indicator to assess regional blood flow by the tracer method is

$$\text{Flow} = \frac{Q/\text{unit time}}{[C_A] - [C_V]} = \frac{Q/\text{unit time}}{[\bar{C}_A]}$$

where Q is quantity of tracer injected, i.e., milligrams, grams, or milliliters; $[C_A] - [C_V]$ is the arterio-venous difference of tracer concentration; and $[\bar{C}_A]$ is the mean arterial concentration that is equal to the equilibrium concentration of the tracer in the blood. Following the injection there is rapid elevation of indicator concentration with subsequent fall-off and ultimate recirculation. By extrapolating to zero, we can calculate the mean concentration of the indicator, in this instance, 4 mg. If the injected tracer amount is 10 mg,

$$\text{Flow} = \frac{10 \text{ mg}/20 \text{ sec}}{4 \text{ mg/cc}} = \frac{1}{8} \text{ cc/sec} = 7.5 \text{ cc/min}$$

over which the integral (or area) was determined. In the example shown, this indicator concentration would be 4 mg/cc. Now if we injected 10 mg of indicator to obtain this curve, the blood flow would equal the quantity per unit time divided by the arterial concentration minus the venous concentration $\left(F = \dfrac{Q_t}{[C_a] - [C_v]} \right)$. Since the initial venous concentration is zero, the blood flow would equal the quantity of tracer per unit time divided by the mean arterial concentration $\left(F = \dfrac{Q_t}{[C_a]} \right)$. Thus, blood flow equals 10 mg per 20 sec divided by 4 mg/cc, that is, 1/8 cc/sec or 7.5 cc/min.

As can be easily seen, this technique requires absolute knowledge of the mean arterial concentration, which is possible when observing dye in the arterial tree; it is, however, impossible when attempting to assess radioactivity concentration externally or to discern arteriographic contrast material concentration by rapid-sequence filming. Even with a spike input of radioactivity into an arterial system, it is not possible to measure blood flow by observing that material with an external counter. *This technique of indicator dilution measures only the mean transit of the material as it progresses down the vascular tree.* This mean transit time t can be defined as the average time it takes a labeled molecule to appear at some distal point where the measurement is being made. Zierler has shown that, when considered in these terms, the mean transit time equals volume divided by flow $(t = V/F)$. Thus, in any system where regional blood volume may change, and where this change cannot be measured, it is not possible to use the transit of a nondiffusible indicator to measure the blood flow. Since the blood volume in most organs can change significantly and nonpredictably, the external detection of a nondiffusible radioactive tracer is usually not a valuable means of measuring blood flow.

6.1.3   Clearance Methods

The second tracer method employed to measure blood flow depends upon the *clearance of materials from the bloodstream* (Fig. 6–4). Saperstein measured the deposition of radioactive potassium as an index of blood flow to the brain. Similarly, Bing has applied clearance measurements to the clinical assessment of myocardial blood flow following the intravenous injection of rubidium. Rubidium and potassium are essentially cleared from the blood in one passage, but they remain in an intercellular location for sufficient time to be detected externally. This clearance may then be used to measure blood flow. Unfortunately, potassium and rubidium are not truly inert, and

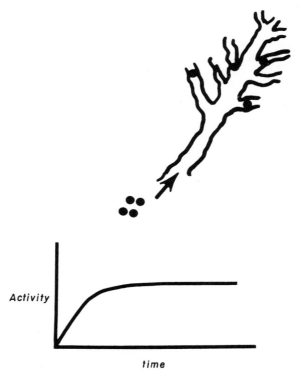

Activity

time

Figure 6–4. Blood flow determined by clearance technique. A popular application of this is to assess the distribution of pulmonary blood flow by particle clearance. However, the standard Saperstein technique, which uses the clearance of potassium or rubidium to measure blood flow, is based on the same tracer principle. Here activity increases until it reaches an early plateau, when the material is fixed within the tissue. The magnitude of this plateau is an index of regional blood flow. To quantitate flow requires knowledge of concentration as it enters into the system. Usually with external detectors this knowledge is not available; therefore, these clearance techniques are useful only in measuring relative regional flow.

their clearance depends upon the viability of the cell, i.e., the integrity of the living cell membrane. Thus, in this instance, a factor in addition to blood flow may be responsible for the data obtained. For example, if diseased myocardial cells tended to exclude the rubidium by the nature of their pathologic process, then the uptake of rubidium would not just measure flow but would combine a measurement of myocardial disease with the flow determination. Since it is frequently difficult to distinguish the diseased muscle cells from normal muscle cells, the uptake of radioactivity does not depend on flow alone, and in pathologic states potassium or rubidium clearance measures blood flow only to viable cells and may not represent a measurement of blood flow to the total tissue.

Unless we know the concentration of the input bolus, clearance techniques are useful only in measuring relative flow, comparing one portion of the tissue with another. Absolute flow determination requires knowledge of radioactivity concentration (e.g., detectable counts per cubic centimeter).

Clearance methods have been popularly applied in clinical medicine to measure regional pulmonary perfusion (Fig. 6–4). Here, one can observe the impact of large (30 to 60 μ) particles in the small ($\sim 8\ \mu$) pulmonary capillary bed and thereby assess the distribution of pulmonary blood flow. It is possible because only a small portion of the capillary bed is blocked, there being very few particles per capillary. This is the basis of clinical lung scanning and provides a useful means of discerning the regional blood flow to the two lungs. It is, however, necessary to know the radioactive concentration as it enters the pulmonary artery if it is wished to quantitate this blood flow. This information is usually not available; therefore the technique is useful only as an index of relative blood flow, comparing one portion of the lung with another. There is no doubt, however, that, if the bolus is well mixed and all particles are stopped in the lung, this clearance method measures only blood flow.

6.1.4   Inert Diffusible Indicators

Recognizing the limitation of nondiffusible-indicator and clearance techniques, many investigators use the Kety-Schmidt application of the Fick principle to quantitate blood flow. Kety and Schmidt initially used the diffusible indicator nitrous oxide N_2O to measure cerebral blood flow. They reasoned that the concentration residing in the brain $[C_B]$ equals the arterial concentration minus the venous concentration $[C_B] = [C_A] - [C_V]$, and the amount of N_2O entering the brain is a product of the flow times the N_2O concentration multiplied by the duration over which the concentration is measured. The more rigorous treatment of these data will be developed later in this chapter (see Fig. 8–3).

Here let us consider the properties that make this diffusible indicator technique work so well for blood-flow measurements. The inert indicator rapidly diffuses throughout the volume of tissue in which flow is being measured. When this can be observed by external counting, i.e., with an inert radioactive gas as the indicator (for example, ^{133}Xe, ^{135}Xe, ^{85}Kr, $^{13}N_2O$), the peak of the curve represents equilibrium throughout the tissue — when the indicator has saturated the tissue (Fig. 6–5). The total volume of tissue being studied is thus defined by the maximum indicator concentration and is independent of a change in regional blood volume. Any subsequent decrease (washout) of the indicator is due to the amount of unlabeled

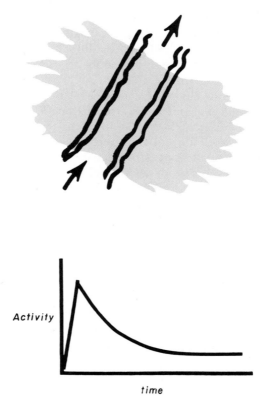

Activity

time

Figure 6–5. The diffusible-indicator approach to measuring blood flow, where the activity reaches a rapid early spike followed by a washout. The rate that the material disappears from the tissue will depend upon the rate that unlabeled blood flows through the tissue and the relative affinity of the blood versus the tissue for this indicator. As long as the material is inert and freely diffusible, this method allows for a quantitative measurement of blood flow, e.g., in cubic centimeters per 100 gm perfused tissue. The volume of indicator distribution in the initial pass is the measure of volume being studied. Flow is calculated from the equation $\bar{f} = \lambda H/A$, where λ equals the partition coefficient, i.e., relative affinity of the tissue versus the blood for the indicator; H equals the maximum height of the curve; and A equals the area under the curve.

blood washing the label out of the tissue. These inert gases tend to be removed rapidly from the circulation by the lungs, thus eliminating the problem of recirculation.

If the relative chemical affinity of blood and tissue for the label is similar, the washout rate is dependent *only* on blood flow (Fig. 6–6). Most indicators, however, do not have identical affinity between blood and tissue, and therefore the *partition coefficient* between blood and tissue must be inserted into the calculation, as a constant, to yield a quantitative measure of blood flow. The partition coefficient is the ratio of

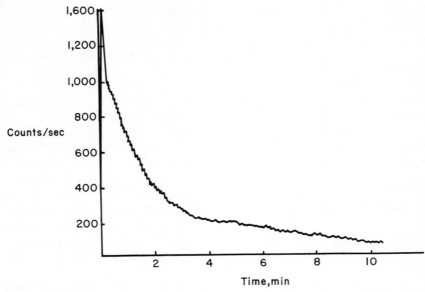

Figure 6-6. Typical washout curve obtained from xenon-133 injection into the brain. Note the rapid rise of the curve and subsequent washout of this inert diffusible indicator, which is not recirculated since it is expelled by the lungs. This typical indicator curve can be analyzed by the formula $\bar{f} = \lambda H/A$. Compartment analysis of these curves is at times useful.

$$\frac{\text{Conc. of tracer in 1 gm of tissue}}{\text{Conc. of tracer in 1 gm of blood}}.$$

Certain indicators, e.g., ^{14}C-antipyrine and ^{15}O-water, do not have a different blood-tissue affinity and therefore are considered to have a partition coefficient of 1, which simplifies the calculation of blood flow.

There are certain biophysical limitations to the inert-gas washout methods that must be recognized. First, the indicator must be freely diffusible so that the rate at which it leaves the vessel is essentially instantaneous with relation to the blood flow that is being measured. In addition to diffusing rapidly, it must also diffuse freely throughout the tissue of interest. These restrictions mean that if the indicator is not delivered homogeneously throughout the tissue, then it will measure the blood flow only in the regions that receive the indicator. The practical limitations of this are obvious; e.g., in the measurement of kidney blood flow in renal cystic disease, if the cyst wall presents a barrier to the indicator, blood flow calculated in cubic centimeters of blood per 100 gm of tissue will measure only the kidney tissue being perfused. However, the same problem would arise if a kidney or heart vessel were completely obstructed and a region beyond the obstruction was not "perfused," in which case diffusion into the region from surrounding vessels might not be instantaneous or might not occur at all. In the former case the inert-gas washout technique would detect only the flow to perfused tissue, and the calculation must be considered as cubic centimeters per 100 gm of *perfused* tissue. However, if diffusion into the area did occur but was delayed, then the rate of diffusion could be detected and could influence the curve, i.e., *become a rate-limiting process* (see Sec. 7.1.2), and the washout would measure not just blood flow but rather an ill-defined combination of abnormal diffusion and blood flow.

A related biological limitation occurs when the external detector observes a rapid transit of radioactivity that does not diffuse. In practical situations the restriction of instantaneous complete diffusion is not always met. For example, when a probe is placed over the temporal lobe of the brain and ^{133}Xe (dissolved in saline) is injected into the internal carotid artery, some of the initial activity detected, contributing to the maximum height of the curve, will not be located within the brain substance. Rather it will be within the bolus of activity traversing the carotid artery, a structure also "seen" by the detector.

In the calculation of blood flow, as discussed later in the book (p. 167), the maximum height of the curve is an important parameter. When a rapid peak in activity—i.e., the so-called "shunt spike" or "non-nutritional flow"—is seen, it is important to recognize the non-diffusible component, which must be corrected to obtain a reliable

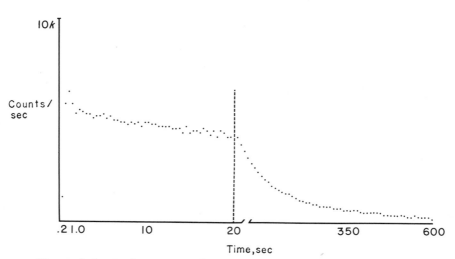

Figure 6–7. Early curve analysis is occasionally necessary to define the accurate maximum height, which here indicates an additional compartment of shunt flow. The first portion (20 sec) of these curves is taken with digital increments at 4/10 sec, and subsequent data are accumulated over 10-sec increments. This amplification of early time is useful to assess early changes in the curve, as described in the text.

measurement of blood flow (Fig. 6–7). Occasionally, rapid (for example, 0.1 sec) sampling durations are necessary to detect this shunt spike. Thus, the instruments must be capable of detecting radioactivity at a sufficiently rapid rate to obtain a statistically valid measurement in a brief time. This time can be defined by the physiological limits of the system under investigation. These are important biological limitations that must be considered when attempting to apply these techniques to quantitate regional blood flow.

Another limitation is related to the physical considerations of the indicator that is being detected. This is not serious when one is observing the N_2O content in vessels (e.g., in a jugular vein) as done by Kety and Schmidt. However, when external detection of a radioisotope is attempted, the volume of tissue seen is dependent upon the efficiency of the detecting system and the energy of the isotope. Lower-energy isotopes (for example, ^{133}Xe) produce scatter so close to the photopeak that the primary photons cannot easily be discriminated from the scatter—resulting in poor anatomic definition. In addition such isotopes emphasize superficial structures since the tissue absorbs a significant fraction of the emitted photons.

When using isotopes with light energy γ-rays (for example, ^{85}Kr and $^{13}N_2O$) the radiation tends to penetrate the collimator (often a lead shield) that is used to define a small area of tissue by excluding photons from adjacent tissue. When sufficient shielding is used to exclude the penetrating photons from these isotopes, the efficiency is decreased and the lead shielding that is used to separate the detecting crystals may obscure adjacent anatomic regions.

Thus, an isotope with a medium energy γ-ray (for example, ^{135}Xe) would appear to be the indicator of choice for these procedures. The only difficulty in using ^{135}Xe is its availability.

When the limitations are appreciated, it becomes apparent that the inert-gas washout techniques (diffusible-indicator methods) are the most reliable tracer methods to measure physiologic blood flow. Therefore, the authors will develop the more rigorous mathematical considerations of such techniques later in this chapter. Now let us define the tracer terms as a prerequisite for more solid consideration of tracer application to volume (distribution) and flow (kinetic) determinations.

6.2   DEFINITION OF TERMS

Tracer kinetics is basically the same whether one is studying a chemical system or a biological system. Terminology, however, differs from one discipline to another. In general, kinetic studies of all kinds involve solving the rate law (or its variations); this is a mathematical statement which relates the rate (change per unit time) at which the particular process occurs to concentrations of various components that are present in the system. For example let us consider the following simple (first order) rate expression:

$$\text{Rate} = k[A]$$

where k is simply a proportionality constant called the *rate constant*, and $[A]$ is the concentration of A, the species we are tracing.

The species A may be a chemical species, such as an atom, a molecule, or an ion; or A may be some other more generalized mass or volume parameter; that is, any discrete region separated from the rest of its environment by boundaries. This more generalized definition of a species is particularly applicable to medical studies.

The following mass or volume parameters have been recommended by the International Commission on Radiologic Units (ICRU):

A *compartment* is an anatomical, physiological, chemical, or physical subdivision of a system throughout which the ratio of concentration of tracer to tracee is uniform. This implies that the rate at which the tracer entering a compartment is mixed with tracee in the compartment is rapid compared with the rate at which the tracer leaves the compartment; for example, a bolus containing macroaggregated particles is thoroughly mixed in the right ventricle prior to its leaving via the pulmonary artery to be distributed throughout the lungs. This distribution is therefore an index of blood flow.

A *space* is an apparent volume—which may also be a compartment—obtained by dividing the amount of retained tracer by the concentration of the tracer at the sampling site. As an example one could use the determination of plasma space. The plasma space is part of the blood pool. The space measured by labeled albumin (at 10 min equilibration) has, in the past, been used erroneously as an index of the plasma space, i.e., plasma volume (see p. 123).

A *pool* is defined as the total amount of substance in a system or subsystem, for example, the blood pool. A pool may, of course, be a compartment; for example, the intravascular albumin pool equals the concentration of albumin times the albumin space; in numbers, 2450 ml × 0.045 gm/ml = 110.25 gm.

If we return to the simple (first-order) rate-constant expression (i.e., Rate = $k[A]$), it is obvious that the concentration in a compartment (e.g., an organ) is not changing; what is changing is the amount of tracer entering or leaving the compartment. This change may then be correlated with a function of the compartment. The term $[A]$ (concentration) must then be an expression which relates the amount of tracer to the size of the compartment. In matrix terms this can be expressed as $a_j Q_j$, where a_j is the specific activity of the tracer in the jth compartment and Q_j is the quantity of tracee in the jth compartment.

6.3 DILUTION TECHNIQUES

Whereas the rate of transfer of a tracer between compartments or pools is a very useful quantity and will be discussed in detail later (p. 130), the *equilibrium value** of the tracer concentration in a single pool is also of great value to determine the physical size of the body pool or space in which the tracer is distributed. Such equilibrium studies are referred to as "isotope dilution" studies. In these studies a known amount of tracer is injected into the pool or space whose size is to be determined. If the tracer has no method of leaving the pool of interest, the egress of activity from the injection site will be as shown in Figure 6.8. The activity diffuses into the whole pool until equilibrium occurs, that is, the state when no further change can be detected. It should be noted that equilibrium is a gross or macroscopic property of a system; on the atomic or molecular level many changes are occurring. After equilibrium has been reached, the concentration is determined. If $[C_1]$ and V_1 are the concentration and volume injected, and $[C_2]$ and V_2 the concentration and volume at equilibrium, then

(6.1) $$[C_1]V_1 = [C_2]V_2$$

(6.2) $$V_2 = \frac{[C_1]V_1}{[C_2]}$$

**Equilibrium value* can be considered to be the amount of tracer observed in a compartment when this amount has become stabilized (i.e., the amount of tracer entering the compartment is equivalent to the amount of tracer leaving the compartment). For practical applications these restrictions are not always strictly enforced in that an equilibrium value may be obtained at a time when the amount of tracer in a compartment is relatively stable when compared with other possible times for measurement.

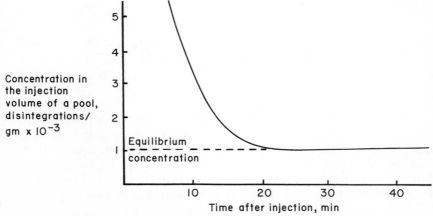

Figure 6–8. The egress of activity from the injection site of a pool as a function of time.

Dilution studies can be performed with dyes (such as Evans blue), stable isotopes, which are detected by mass differences (such as deuterated water), and radioactive isotopes (such as tritiated water, sodium-24, potassium-24, and chromium-51). The major concern in studies of this type is that the tracer stay in the volume of interest for the duration of the observation; in other words, the transfer of label out of the volume during the time required to reach equilibrium must be small.

6.3.1 Red-Cell Volume

An example of where the "equilibrium stability" becomes important is in the determination of plasma volume. Obviously, at any moment in time there is a finite (absolute) volume occupying the intravascular space. This whole-blood volume is made up of a *red-cell volume* and a *plasma volume*. The red-cell volume can be determined by labeling a red-cell sample A with ^{51}Cr, injecting it intravenously, and removing another sample B from a vein remote from the injection site at equilibrium, usually 10 to 20 min after injection. The radioactivity can be expressed as counts per minute as determined in a standard well counter. In this case the concentration injected (C_1) equals the counts per minute obtained from a 1 ml aliquot of the labeled red cells. The concentration removed (C_2) equals the counts per minute obtained from sample B. The formula $C_1V_1 = C_2V_2$ can be expressed in this instance as counts per minute per milliliter injected times milliliters injected equals counts per minute per milliliter removed times the volume of distribution (i.e., the red-cell volume). Since counts per minute per milliliter injected times milliliters injected can be expressed as counts injected, the red-cell volume can simply be determined by

$$\text{Red-cell volume} = \frac{^{51}\text{Cr counts per minute injected}}{^{51}\text{Cr counts per minute removed}}$$
$$\times \text{ ml red cells removed.}$$

Although there are subtle problems involved in this measurement, such as the elution of label from the cells and the inaccuracy of hematocrit measurement, this technique represents an almost ideal isotope dilution method, because there is usually rapid and complete mixing of red-blood cells. Also once equilibration is reached, it is relatively stable for the duration of measurement. The red cells are contained within a rather homogeneous space* and they do not tend to shift from one equilibrium residence to another. This sharply defined "border" of a "space" is termed *holostic integrity* and the red-cell volume determination is said to have a high degree of holostic integrity.

6.3.2 Plasma Volume

When measuring the plasma volume with either Evans blue dye or radioactive iodine tagged to albumin, the situation is very different from that discussed for red-cell volume. Although there is a finite intravascular plasma volume, it is extremely difficult to find a label that homogeneously distributes throughout it and at the same time does not "leak" into the surrounding interstitial space. Although radioactive albumin is rather easily tagged and assayed, it represents a poor label for the determination of the plasma volume. In fact, the "true" or "absolute" plasma volume cannot be determined by using radioactive albumin, because intravascular albumin is continuously leaving the intravascular space and entering the extravascular space (compartment). In normal man approximately 40 per cent of the total body albumin is located extravascularly. This percentage decreases with exercise and increases with bed rest.

The rate at which the transcapillary transport of albumin takes place varies considerably in different organs. It occurs rapidly in the liver but requires a matter of hours in the lung. By observing the intravascular albumin label over a prolonged period of time (hours), we can see the volume of distribution (albumin space) increase as more and more of the label leaves the intravascular space (Fig. 6–9). In a longer time (days) a total body equilibrium of albumin will be reached, representing the combined intra- and extravascular space. The reliable determination of this total body albumin space is complicated by the loss of the label due to catabolism of the albumin. This loss of the label must be taken into account when attempting to measure the total albumin space. The rate of catabolism can be inferred by observing the plasma radioactivity for a sufficiently long period (e.g., 20 days) so that the rate of change in plasma activity is due only to catabolism and not due to further equilibration. This change is called the terminal slope of the plasma albumin curve. If

---

*We acknowledge the "extravascular residence" of red cells in the spleen. However, this is not a significant factor in humans, where the spleen does not tend to "contract." Splenic contraction could release a different population of red cells (i.e., those with a different concentration of labeled cells than the circulating blood) into the circulation. In other animals, such as dogs, uncontrolled splenic contraction may significantly alter the red-cell mass measurement.

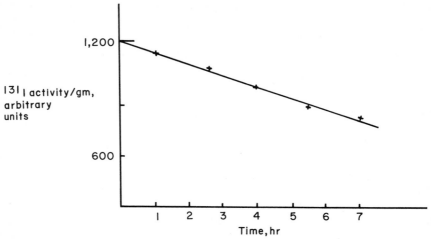

Figure 6–9. Graphical method of determining equilibrium size of a space if tracer is leaving the space by catabolism.

catabolism has been constant for the duration of the study, then the total body albumin space can be determined by extrapolating the "terminal slope" back to the "early portion" of the curve and correcting for the label loss due to catabolism. We acknowledge the additional problems of inadequate labeling procedures resulting in some free iodine as well as recirculation of iodine liberated from the albumin by catabolism. However, when all these factors are taken into account, it appears that one may obtain an equilibrated measurement of albumin space — but this is not equivalent to measuring the plasma volume.

Prior to systemic equilibration the albumin space determination changes with time; for example, about 8.5 per cent of the label leaves the intravascular space each hour in normal man. Although this transcapillary transport rate is rather stable in normal man, in many pathologic states it can vary remarkably, thereby limiting the utility of albumin-space determination as an absolute measurement of plasma volume.

For practical purposes the 10 min albumin space (where the sample is removed 10 min after the label is injected) has been assumed to be representative of the plasma volume. As pointed out above, this assumption is beset with problems, even when limited only to a comparison of 10 min albumin spaces as indices of the plasma volume in different individuals. For example, if the liver enlarges, as in congestive heart failure, the rapidly equilibrating hepatic extravascular albumin space can contribute disproportionately to the plasma-volume determination, and the plasma volume may appear larger without an increase in the finite (absolute) intravascular volume. It is, therefore, preferable to speak in terms of the albumin space at a time after injection (e.g., 10 min albumin space) rather than to infer that this measurement is an index of plasma volume. The 10 min albumin space may be an extremely useful determination as long as we recognize what we are measuring. The early albumin space has a poor holostic integrity.

Other plasma-volume labels have been advocated to offset some of the above problems. To date, labeled gamma globulin ^{125}I-IGg appears to have the greatest holostic integrity as a label of measure of the absolute plasma volume. Although some problems exist with this label, the fact that it tends to remain within the blood pool is a distinct advantage when compared with albumin. The apparent space indicated by gamma globulin after ten minutes is 90 per cent of the apparent space indicated using labeled albumin. The reason for this discrepancy is that the albumin determination includes some tracer that has already left the intravascular pool.

6.3.3   Total Body Water

Total body water and red-blood cell volume are the dilution techniques with the greatest holostic integrity; i.e., they equilibrate into a rather stable volume with only a small transfer out of this volume. In determining total body water, either deuterated or tritiated water is injected into the subject, and Equation 6.2 (p. 121) can be altered to

(6.3)
$$V_2 = \frac{V_1[C_1] - V_u[C_u]}{[C_2]}$$

where V_1, $[C_1]$, V_2, and $[C_2]$ have the same meaning as in Equation 6.2, but $[C_u]$ and V_u are the concentration of the tracer in and the volume of the urine excreted between injection and equilibrium. In this case equilibrium time is less than 6 hr, during which period the average adult loses 0.1 l of water ($\simeq 0.2$ per cent of the body water). In general, then, if the urinary loss is negligible, the equation is simply as shown before (Equation 6.2).

$$V_2 = \frac{V_1[C_1]}{[C_2]}$$

For determination of body water as discussed previously, one can use either a radioactive or a nonradioactive tracer. In the case where deuterated water is used, concentration will be measured in grams per cubic centimeter. On the other hand, for a radioactive tracer it is usual to express this concentration in disintegrations or standard counts per cubic centimeter.

6.3.4   Exchangeable Sodium

The situation for the determination of body sodium space presents a different problem. Radioactive sodium (for example, ^{22}Na) is used to determine the body sodium space. However, almost 25 per cent of total body sodium is in bone and exchanges very slowly with the injected sodium. Ultimate equilibration may not occur or at least does not occur until long after the isotope has decayed; the isotope dilution technique is, therefore, limited to measuring exchangeable sodium (Na_{ex}) rather than total body sodium, i.e., it measures total body sodium minus nonexchangeable sodium.

6.3.5   Additional Techniques of Dilution Studies

Many other techniques have been developed and applied, but a detailed treatment of their limits and applications is beyond the scope of this book. The aforementioned examples of dilution studies should serve to alert the reader to the general techniques and some common problems encountered in their application.

Thus, dilution techniques are useful in experimental biology and clinical medicine and may be applied to any system where the dilution of a tracer in the initial compartment or space is fast compared with its transfer to other compartments or spaces. It is most important to recognize their limitations in order to use such techniques properly.

BIBLIOGRAPHY

The bibliography for Chapter 6 appears at the end of Chapter 8.

Chapter 7

MATHEMATICAL DEVELOPMENT OF TRACER KINETICS

7.1 KINETIC ANALYSIS

Kinetic analysis is the process of detailing the transition of a system from one state to another. In this chapter mathematical solutions will be applied to model systems that approximate complex biomedical systems. Such techniques can be used to greatly simplify the biological applications of the tracer technique.

7.1.1 Input Function

Unlike a chemical reaction, in which initially all the tracer may be placed in the system as one species and then its transfer to other species observed, a biological system may not be receptive to this kind of procedure. The tracer initially may not be entirely in one compartment. In a chemical system, for example, a reaction which has been widely studied is the rate at which a ferric ion changes to a ferrous ion. The reaction may be followed by using tracer ^{55}Fe or ^{59}Fe. All the radioactivity may initially be in the form of ferric ion and its rate of transit studied.

In vitro studies of biological reactions may also be carried out as chemical reactions. The in vivo analogy to the chemical and in vitro

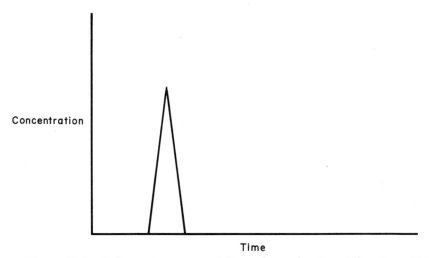

Figure 7-1. Schematic representation of a pulse input function. For a true pulse input the time scale should be microscopic.

situations would be a single injected dose that is instantaneously absorbed by a compartment of interest. Unfortunately, there are many biological processes that do not fit this picture, and for this reason it is extremely important to study the way in which the tracer enters the compartment of the system. Descriptions of the modes of input are called *input functions, which are simply mathematical statements describing the amount of material entering a system or compartment of a system as a function of time.* In other words, an input function defines the rate at which a substance enters a system or compartment.

If one were to represent a chemical or in vitro input function, it would simply be a spike, as shown in Figure 7-1. Pulse input functions can be observed, however, in in vivo studies by arterial injection, similar to the technique discussed previously with inert-gas washout studies to measure cerebral blood flow. The tracer dissolved in saline is injected into the carotid artery of the subject and arrives at the brain in a sharp spike. Another similar example may occur with studies of body kinetics, when the diffusion of the tracer into the blood or plasma pool is very fast compared with its exchange with other compartments. When all the tracer is initially distributed randomly in one compartment, the biological system is a direct analogy to the chemical system, when the tracer is initially distributed randomly in one chemical compound.

A common input function in cell kinetics is known as a *cohort function* and is shown in Figure 7-2. The shape of this curve should be very familiar, as it approximates a normal distribution curve. What is meant, then, by stating that a particular system has a cohort input

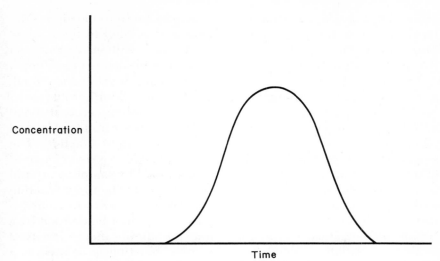

Figure 7–2. A cohort input function.

function? Simply that the tracer enters the system in a finite time during which the amount of tracer entering distributes itself normally; that is, at some time t the maximum of the tracer input function enters the system as shown in Figure 7–2.

A system that exhibits a cohort input function is the study of cell renewal patterns. Not all the cells are in the same phase of their cycle. As DNA is synthesized during only one phase, introduction of the DNA precursor thymidine, labeled with tritium, will label only cells in this phase. If the labeling is for a period of time that is short compared with the length of one cycle, the input of labeled cells entering the cycle will be cohort input.

Another common input function is an *exponential function.* In most cases the input is exponential simply because entrance of the tracer into a compartment of interest is dependent upon its rate of disappearance from some initial compartment. Consider, for example, iodohippuran techniques to measure renal plasma flow. Iodohippuran is injected intravenously and equilibrates between the plasma and interstitial fluid. The entrance of iodohippuran into the kidney and its subsequent excretion, as measured by an external detector, depend on the release of the tracer from this combined plasma and interstitial space. Input of tracer into the kidney is exponential, whereas input of the isotope into the series of spaces (compartments) is a pulse or spike. In order to have a pulse input into the kidney, one must inject iodohippuran directly into the renal artery.

Ferrokinetic studies (see Sec. 8.1.3) represent an example of when the input functions into various pools from the plasma pool are exponential. Again, an initial tracer dose may be introduced "instantaneously" into the plasma. The appearance of label in the marrow iron transit pool or extravascular labile pool will be exponential because the disappearance of tracer from the plasma is exponential.

Input functions may be more complex than cohort or exponential, as described in the introduction to this chapter and further discussed in Section 8.2.1. Kety and Schmidt initially studied cerebral blood flow by using nitrous oxide, not injected into the carotid artery but inhaled; this system exhibits a very complex input function. The subject inhales oxygen containing a certain per cent N_2O over a period of time. Input of N_2O to the brain is determined by sampling arterial blood and measuring the amount of N_2O present. The input function will depend upon the uptake of the tracer into a series of compartments, namely, most of the body organs, and cannot be defined by a simple mathematical expression. From the simple theory involved in the calculation of blood flow, the input function is defined as a function of time and the blood flow is derived from the difference between input and output function. Complex input functions may require computer programs for rigorous solutions, although even with such sophisticated treatment only an approximation can be obtained.

Another example of a complex input function occurs in some leukokinetic studies in which the input functions, normally of the cohort type, are altered in order to verify the model proposed.

In planning any tracer experiment, it is necessary to consider the type of input functions one is using. Analyses of complex systems, which are discussed later in this book (p. 166) are much easier if there is a pulse-type input. Therefore, if at all possible it is preferable to design a tracer experiment with a pulse input.

7.1.2   Compartmental Analysis

As has been previously mentioned, compartmental analysis is analogous to chemical kinetic analysis. The reaction rate is a function of the concentration of some or all components of a system and may be expressed as a derivative. Consider the reaction

$$A + B \rightarrow C$$

The rate at which A disappears is expressed

$$-\frac{d[A]}{dt} = \text{time rate of disappearance of } A$$

where t is time and $[A]$ the concentration of A. The negative sign in the equation indicates that the concentration of A decreases with increasing reaction time and ensures that the rate will be numerically positive. The rate may also be written

$$\frac{-d[B]}{dt} \quad \text{or} \quad \frac{+d[C]}{dt}$$

in terms of the disappearance of B or formation of C. The rate law of a reaction is usually determined experimentally because it is not feasible to predict it from theoretical considerations except for the most simple reactions. In general, it is also not possible to predict the rate law from knowledge of the overall reaction, because almost all chemical reactions are complicated and usually consist of a series of simple steps; that is, it is not always obvious from the overall reaction which species affect the rate of the reaction.

The rate law is a mathematical expression of the relationship between the rate and the concentrations of various components in the system; for example, as described earlier the rate may be proportional to the concentration of one component:

$$\frac{-d[A]}{dt} = k[A]$$

This relationship is not limited to the first power of the concentration. In fact, a more general statement of the rate law is

(7.1)
$$-\frac{d[C_1]}{dt} = k[C_1]^{n_1}[C_2]^{n_2}[C_3]^{n_3} \cdots$$

where $d[C_1]/dt$ is the rate of disappearance of component $[C_1]$; $[C_1]$, $[C_2]$, $[C_3]$, ... are the various components on which the rate depends; and where n_1, n_2, n_3, ... are the various exponents. The *order of the reaction* is defined as N, where

(7.2)
$$N = n_1 + n_2 + n_3 + \cdots$$

If the rate is proportional only to the concentration of one species raised to the first power ($N = 1$), the reaction is termed a *first-order* reaction. It is possible for $N = 0$; this reaction is termed a *zero-order* reaction. For $N = 2$, the reaction is *second order*. Zero- and first-order reactions will be discussed separately in the following text, as most biological processes resemble these categories.

A series of elementary reactions usually make up the overall reaction and, when taken together, are known as the mechanism of the reaction. The overall reaction rate in such a scheme will often be fixed by the rate of the slowest elementary reaction. This "slow step" is called the *rate-determining step*. The form of the rate law will then reveal what species or components take part in the rate-determining step. It should be clear that there may be several mechanisms consistent with the experimental rate law. Many kinetic studies are concerned with the confirmation of one mechanism over another by such techniques as the detection of proposed intermediates in the mechanism.

There are many factors which may affect the rate of a reaction. In addition to concentration, the physical state of the reactants, temperature, pressure, and catalysts may all contribute. Usually increasing temperature and pressure for a gaseous reaction will increase the rate. A suitable catalyst may also increase the rate of a reaction. A catalyst is, of course, a substance that accelerates the speed of a reaction but is recovered unchanged chemically at the end of the reaction. Most mechanisms for reactions involving a catalyst suggest that the catalyst is involved in some intermediate step and is then regenerated. The catalyst permits the reaction to go by some different path, one of lower energy than if the catalyst were not present. The most important group of biological catalysts are enzymes, which include some of the most efficient catalysts known. It should be noted that it is possible to have a negative catalyst, or inhibitor, which will decrease the rate of the reaction.

7.1.3 First-Order Reactions

If a reaction is first-order, then the rate is dependent upon the concentration of one of the components in the system under study. The rate expression is

(7.3)
$$-\frac{dC}{dt} = k[C]$$

A plot of concentration versus time for a first-order reaction, shown in Figure 7–3, is nonlinear; that is, the rate changes throughout the reaction. The rate of a reaction is usually not obtained directly from experiment; in most cases the experimentally observed quantity is concentration as a function of time. It is therefore more convenient to work with the integrated form of the rate law. Rearrangement of the rate-law equation gives

$$-\frac{dC}{C} = k\,dt$$

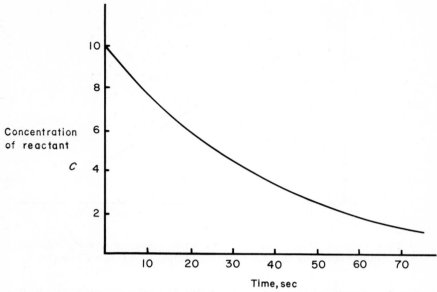

Figure 7–3. A plot of concentration versus time for a first-order change. It should be noted that the shape of this is the same as that given for radioactive decay in Figure 2–5.

and integrating both sides of the equation gives

$$(7.4) \qquad -\int_{C_0}^{C} \frac{dC}{C} = k \int_{0}^{t} dt$$

where C_0 is the initial concentration and C is the concentration at some time t after the start of the reaction. The limits of integration for time are $t = 0$ to $t = t$. The result is

$$(7.5) \qquad \ln \frac{C_0}{C} = kt$$

or

$$(7.6) \qquad \ln C = \ln C_0 - kt$$

or, for log base 10,

$$(7.7) \qquad \log C = \log C_0 - \frac{kt}{2.303}$$

or by simply rearranging Equation 7.6 for log e,

$$(7.8) \qquad C = C_0 e^{-kt}$$

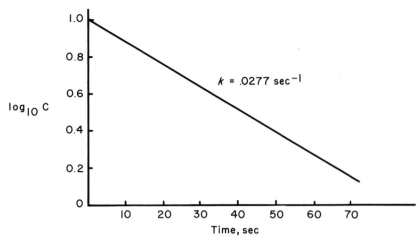

Figure 7–4. A plot of $\log_{10}$ concentration versus time for a first-order change. The k (rate constant) calculated from this plot is 0.0277 sec^{-1}.

For a first-order reaction, a plot of log C versus time will yield a straight line whose slope is $-k/2.303$ and whose y intercept is log C_0. The rate constant k may be easily determined in this way. The data from Figure 7–3, if plotted this way, give a value of $k = 0.0277$ sec^{-1} (Fig. 7–4).

There exists an interesting experimental condition whereby a reaction may appear to follow first-order kinetics but, in fact, is a pseudo first-order reaction. Assume the concentration of a reactant is so large that it does not change appreciably during the course of the reaction. It may then be assumed constant throughout the reaction. Take, for example, the elementary reaction

$$A + B \rightarrow C$$

which is a second-order reaction whose rate is

(7.9) $$-\frac{dA}{dt} = -\frac{dB}{dt} = k[A][B]$$

If the concentration of A is very large, essentially constant, and equal to $[A_0]$, the initial concentration, then the rate of the reaction is

(7.10) $$-\frac{dB}{dt} = k[A_0][B] = k'[B]$$

where k' is a pseudo first-order rate constant formed from the product of the true second-order rate constant and $[A_0]$.

As has been indicated, concentration is the usual variable that is measured. Concentration is often used instead of the amount of substance present because it is an intensive property, or independent of the size of the system. There are, however, variables other than concentration that are used in defining the rate of a reaction. In gaseous reactions, pressure is used. In racemization, optical rotation may be used. These variables are usually linearly related to concentration and so may be treated as equivalent to concentration. For tracer studies, it has already been shown that the variable used is equal to (again in matrix notation) $a_j Q_j$, where a is the specific activity of the tracer and Q_j the amount of tracee present in the jth compartment, because the concentration of tracee within the compartment is not changing, but the amount of tracer present is.

The following scheme represents the exit of a tracer from compartment A to compartment B with first-order kinetics:

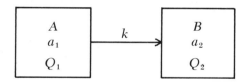

where $a_1 Q_1$ is the amount of tracer in compartment A, and $a_2 Q_2$ is the amount of tracer in compartment B. The rate of tracer exit from A equals the rate of entrance into B, or

(7.11) $$-\frac{da_1 Q_1}{dt} = \frac{da_2 Q_2}{dt} = ka_1 Q_1$$

The rate is dependent upon the amount of tracer present; in other words, the compartmental scheme is completely analogous to a chemical reaction.

From the integrated form of the rate law for a first-order reaction, one can easily deduce that the nonlinear relationship obtained by plotting C versus t is simply an exponential curve. The rate of exit of tracer from compartment A equals the rate of entrance into compartment B and both are exponential functions. Clearly, the observation of an exponential input function for some compartments depends in most cases upon a first-order reaction of the tracer leaving some initial compartment.

As will be demonstrated later, most reactions in a compartmental scheme are considered first order. Specific biological examples of first-order reactions will also be presented.

7.1.4   Zero-Order Reactions

If the order of a reaction is zero, then the rate is independent of the concentration of components in the system under study and remains constant throughout the reaction. The rate law for this case is given by

(7.12)
$$-\frac{dC}{dt} = k[C_0] = k = \text{constant}$$

Again one may integrate the equation to give an expression in terms of concentration rather than rate. Rearrangement yields

(7.13)
$$-\int_{C_0}^{C} dC = k \int_0^t dt$$

where the limits of integration are the same as defined previously. The result is

(7.14)
$$C_0 - C = kt$$

or

$$C = -kt + C_0$$

A plot of concentration versus time, as shown in Figure 7–5, will yield a straight line whose slope is equal to $-k$ and whose y intercept is equal to C_0.

As with first-order reactions, there are conditions in which a reaction appears to follow zero-order kinetics but is actually a pseudo zero-order reaction. Consider the elementary reaction

$$A \rightarrow B$$

in which the rate is

(7.15)
$$\frac{-dA}{dt} = \frac{dB}{dt} = k[A]$$

that is, it is a first-order reaction. Suppose, again, that A is very large and is essentially constant. The product B, however, is still forming, and the rate of the reaction is

(7.16)
$$\frac{dB}{dt} = k[A_0] = k'$$

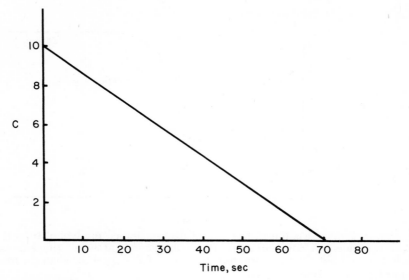

Figure 7–5. A plot of concentration versus time for a zero-order change. The zero-order rate constant $k = 0.143$.

where A_0 is the initial concentration of A, which, when combined with the true first-order rate constant, gives a pseudo zero-order rate constant for the reaction.

Let us examine the situation for compartmental analysis when a zero-order reaction is observed. Consider the exit of tracer from compartment A to compartment B with zero-order kinetics. The rate expression is

(7.17)
$$- \frac{da_1 Q_1}{dt} = \frac{da_2 Q_2}{dt} = k = \text{constant}$$

or the rate of exit and entrance is constant.

One of the best known biological reactions to exhibit zero-order kinetics is the clearance of ethanol from the blood. The rate at which the ethanol is cleared from the blood is essentially independent of the blood ethanol concentrations; that is,

$$- \frac{d[\text{alcohol}]}{dt} = \text{constant}$$

This reaction, like most enzymatic reactions, is zero order only when the enzyme, in this case ethanol dehydrogenase, is saturated. At low ethanol concentrations the amount of reaction is proportional to the amount of ethanol present, shown in Figure 7–6; i.e., the reaction is

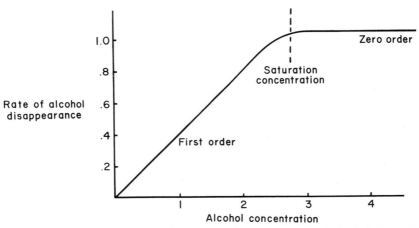

Figure 7–6. Variation in the rate of disappearance of alcohol from the blood with alcohol concentration. It can be seen that when the enzyme alcohol dehydrogenase is saturated, the rate of the reaction changes from first order to zero order.

first order. The ethanol concentration has to be at 0.1 mg/gm blood or below for the metabolic process to appear first-order (after drinking a fraction of a highball, the rate becomes zero-order). This limiting concentration occurs because all available enzyme is used to metabolize ethanol. When this state is reached, no matter how much more ethanol is added to the system, the amount metabolized cannot increase, as there is no free enzyme available. The complete rate laws for clearance of ethanol are, therefore,

$$\frac{-d[\text{alcohol}]}{dt} = k[\text{alcohol}]$$

at low ethanol concentration;

$$\frac{-d[\text{alcohol}]}{dt} = k = \text{constant}$$

at high ethanol concentration.

All enzymatic reactions and, indeed, all catalytic reactions exhibit this change in order.

7.1.5   Second–Order Reactions

Second-order reactions are quite common chemically. Most biological reactions, on the other hand, can usually be handled as zero order or first order. The simplest second-order reactions are

$$A + B \rightarrow C$$

where

$$\text{Rate} = k[A][B]$$

or

$$2A \rightarrow B$$

where

$$\text{Rate} = k[A]^2$$

For the latter case, the integrated form of the rate expression is simply

(7.18)
$$\frac{1}{[A]} - \frac{1}{[A_0]} = kt$$

where $[A_0]$ is the initial concentration, and $[A]$ is the concentration at some time t after the start of the reaction.

The integrated form of the former situation (Rate $= k[A][B]$) is not so straightforward, but the result may be simplified to

(7.19)
$$\frac{1}{B_0 - A_0} \ln \frac{A_0 B}{B_0 A} = kt$$

where the A_0, A, B_0, B have the usual meaning.

An alternative, but equivalent, way of expressing the rate law simplifies the determination of k for a second-order reaction. Let x be the concentration of reactant that has reacted in time t. If A_0 is the initial concentration of A, then $[A_0 - x]$ is the concentration of A remaining at time t. Similarly, for B, $[B_0 - x]$ is the concentration of B remaining at time t. The rate law in terms of x is

(7.20)
$$\frac{dx}{dt} = k[A_0 - x][B_0 - x]$$

and the integrated form is

$$\frac{1}{[A_0 - B_0]} = \ln \frac{B_0[A_0 - x]}{A_0[B_0 - x]} = kt$$

A plot of log $[B_0(A_0 - x)/A_0(B_0 - x)]$ versus time will result in a straight line whose slope is $k(A_0 - B_0)/2.303$, from which k may be calculated.

7.1.6 Autocatalytic Reactions

In certain types of changes of the general form

$$A \rightarrow B$$

the product formed (B) catalyzes the reaction. A biochemical example of a reaction of this type is the conversion of trypsinogen (A) into trypsin (B), where the trypsin catalyzes the reaction.

In this case the rate expression is

(7.21)
$$-\frac{dA}{dt} = kAB$$

As A is simply converted into B, the amount of A and B present at any time t is equal to the sum of the initial concentrations of A and B $(A_0$ and $B_0)$.

Therefore,

$$A_0 + B_0 = A + B$$

$$B = A_0 + B_0 - A$$

The rate equation is, therefore,

(7.22)
$$-\frac{dA}{dt} = kA (A_0 + B_0 - A)$$

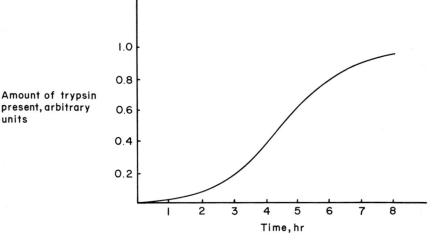

Figure 7-7. A sigmoid-type curve showing the amount of trypsin formed from trypsinogen.

and integration of this equation gives

(7.23) $$\frac{1}{A_0 + B_0} \ \ln \ \frac{A_0 B}{B_0 A} = kt$$

The formation of trypsin (B) with time occurs as shown in Figure 7–7 and it should be noted that without the presence of a trace of trypsin (B_0) initially, the reaction would not occur.

This type of autocatalytic growth curve is met not only in tracer kinetics; population growth shows the same trends, as well as the characteristic "sigmoid" curve.

7.1.7 Consecutive and Concurrent First-Order Changes

Systems of interest often contain consecutive first-order changes. Typical examples of consecutive changes are the decay of a radioactive atom to a stable atom via a series of radioactive daughter atoms, for example,

$$^{99}\text{Mo} \overset{k_1}{\to} {}^{99\text{m}}\text{Tc} \overset{k_2}{\to} {}^{99}\text{Tc}$$

Such generator systems are used extensively in nuclear medicine to produce short-lived tracers.

The general type of consecutive first-order change is

$$A \overset{k_1}{\to} B \overset{k_2}{\to} C \overset{k_3}{\to} D, \text{ etc.}$$

but we shall solve only for the type of system where there are three components.

$$A \overset{k_1}{\to} B \overset{k_2}{\to} C$$

The rate of disappearance of A is

(7.24) $$\frac{-dA}{dt} = k_1[A]$$

whereas the rates of formation of B and C are

(7.25) $$\frac{dB}{dt} = k_1[A] - k_2[B]$$

and

(7.26)
$$\frac{dC}{dt} = k_2[B]$$

If initially there is a concentration A_0 of A and zero of B and C, then at time t, $A + B + C = A_0$. By integration of Equation (6.27), $A = A_0 e^{-k_1 t}$; and by substitution in Equation (7.25),

(7.27)
$$\frac{dB}{dt} = k_1 A_0 e^{-k_1 t} - k_2 B$$

This equation $dB/dt + k_2 B = k_1 A_0 e^{-k_1 t}$ is a differential equation, the solution for which follows:

(7.28)
$$B = \frac{k_1}{k_2 - k_1} A_0 (e^{-k_1 t} - e^{-k_2 t})$$

As we know the equations for the variation of A and B with time, we may substitute these values into $A + B + C = A_0$; the result yields

(7.29)
$$A_0 e^{-k_1 t} + \frac{k_1}{k_2 - k_1} A_0 (e^{-k_1 t} - e^{-k_2 t}) + C = A_0$$

and solving for C,

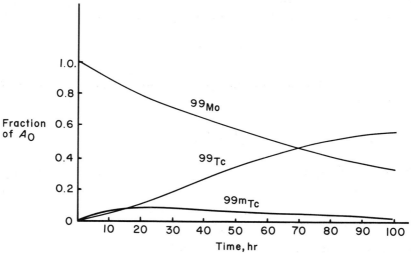

Figure 7–8. Variation in amounts of ⁹⁹Mo, ⁹⁹ᵐTc, and ⁹⁹Tc in a system initially containing all ⁹⁹Mo. It should be noted that we are plotting the fraction of molecules present in each isotopic form, not the number of radioactive disintegrations observed.

(7.30)
$$C = A_0 \left[1 - \frac{(k_2 e^{-k_1 t}) - (k_1 e^{-k_2 t})}{k_2 - k_1} \right]$$

With this technique, the solutions can be calculated for any number of consecutive changes. For the three-component system

$$^{99}\text{Mo} \xrightarrow{k_1} {}^{99\text{m}}\text{Tc} \xrightarrow{k_2} {}^{99}\text{Tc}$$

where $k_1 = 1.04 \times 10^{-2}$ hr^{-1} and $k_2 = 1.15 \times 10^{-1}$ hr^{-1}. The variation in the concentrations of the three isotopes with time is shown in Figure 7–8.

Concurrent first-order changes occur in many washout-type studies. For example, as discussed previously, the washout of inert gas from the kidney appears as if the tracer enters several compartments and is washed out from these compartments separately. This may be represented as follows:

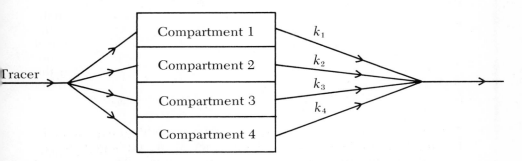

The chemical analog would be

$$A \xrightarrow{k_1} E$$

$$B \xrightarrow{k_2} E$$

$$C \xrightarrow{k_3} E$$

$$D \xrightarrow{k_4} E$$

that is, four parallel first-order reactions with a common product. The rates of disappearance of A, B, C, and D are simply

(7.31)
$$-\frac{dA}{dt} = k_1 [A] \qquad -\frac{dB}{dt} = k_2 [B] \qquad \cdots$$

These rate equations integrate to

(7.32) $$[A] = [A_0]e^{-k_1 t} \qquad [B] = [B_0]e^{-k_2 t} \qquad \ldots$$

where $[A_0]$, $[B_0]$, and so forth, are initial concentrations. At the end of the reaction, that is, at $t = \infty$, the concentration of E will equal $[A_0] + [B_0] + [C_0] + [D_0]$, or

$$[E_\infty] = [A_0] + [B_0] + [C_0] + [D_0]$$

At any time t, the concentration of E is

(7.33) $$E = (A_0 - A) + (B_0 - B) + (C_0 - C) + (D_0 - D)$$

but

$$A = A_0 e^{-k_1 t} \qquad B = B_0 e^{-k_2 t} \qquad \ldots$$

(7.34) $$E = A_0 - A_0 e^{-k_1 t} + B_0 - B_0 e^{-k_2 t}$$

and

$$E_\infty = A_0 + B_0 + C_0 + D_0$$

Therefore,

(7.35) $$\Delta E = E_\infty - E = A_0 e^{-k_1 t} + B_0 e^{-k_2 t} + C_0 e^{-k_3 t} + D_0 e^{-k_4 t}$$

The change in E is equal to the sum of the individual exponentials. For inert-gas washout from the kidney, let A_1, A_2, A_3, and A_4 be the sizes of the compartments 1, 2, 3, and 4, respectively. The variation of total tracer concentration A in the compartments with time is obtained by using an external counter to measure the activity in the kidney as a function of time. Theoretically this change in activity with time is given by

(7.36) $$A = A_1 e^{-k_1 t} + A_2 e^{-k_2 t} + A_3 e^{-k_3 t} + A_4 e^{-k_4 t}$$

A curve consisting of three components can be analyzed by plotting on semilogarithmic paper and "peeling" off the components with the longer half-lives. In the example shown in Figure 7–9 the curve is made up of three compartments, and the relative sizes are shown. This technique, although much used for interpreting medical data, should be employed with caution if the systems appear to contain more than three compartments. The experimental error in points at large t are often greater than when t is small. An error in the calculation of the last compartment will lead to errors in the preceding compartments.

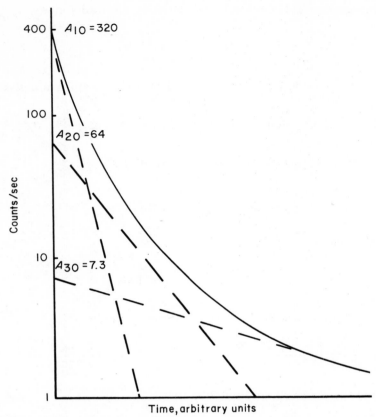

Figure 7–9. Three exponential decays have been "peeled" from the complex function shown plotted on semilog paper.

7.1.8   Fractional-Life Method

Another commonly used method to determine rate constants is the fractional-life method, the most common example of which is the half-life of a reaction. To define again, the half-life of a reaction is the time required for half of a given reactant to be used up. The half-life method is applicable to reactions in which the rate expression has the following form:

(7.37) $$\frac{dx}{dt} = k(A_0 - x)^n$$

where A_0 is the initial concentration of reactant, x the amount of reactant used after time t, and n the order of the reaction.

Integration of Equation (7.37) between the limits gives

(7.38)
$$\int_0^x \frac{dx}{(A_0 - x)^n} = k \int_0^t dt$$

For the case of $n = 1$, the integral is a logarithmic expression given as

(7.39)
$$[\ln (A_0 - x)]_0^x = \ln A_0 - \ln (A_0 - x) = \ln \frac{A_0}{A_0 - x} = kt$$

At $t_{1/2}$, $x = A_0/2$ (one-half A has reacted); therefore,

(7.40)
$$\ln \frac{A_0}{A_0 - A/2} = \ln 2 = 0.693 = kt_{1/2}$$

One obtains

(7.41)
$$k = \frac{0.693}{t_{1/2}} \qquad \text{for } n = 1$$

For those cases of $n \neq 1$, the integration gives

(7.42)
$$\frac{1}{n-1} \left[\frac{1}{(A_0 - x)^{n-1}} \right]_0^x = \frac{1}{n-1} \left[\frac{1}{(A_0 - x)^{n-1}} - \frac{1}{A_0^{n-1}} \right] = kt$$

Again at $t_{1/2}$, $x = A_0/2$, and one obtains

(7.43)
$$\frac{1}{n-1} \left[\frac{1}{(A_0/2)^{n-1}} - \frac{1}{A_0^{n-1}} \right] = \frac{1}{n-1} \frac{2^{n-1} - 1}{A_0^{n-1}} = kt_{1/2} \qquad n \neq 1$$

For a second-order reaction, $n = 2$,

(7.44)
$$t_{1/2} = \frac{1}{kA_0}$$

or

$$k = \frac{1}{A_0 t_{1/2}}$$

For a third-order reaction, $n = 3$,

(7.45)
$$t_{1/2} = \frac{3}{2kA_0^2}$$

or

$$k = \frac{3}{2A_0^2 t_{1/2}}$$

A common application of half-life is with regard to radioactive isotopes, where $t_{1/2}$ is given as a characteristic physical property of the isotope. Radioactive decay is a first-order reaction, so that k (the rate constant for decay) is simply calculated from Equation (7.41). Important usage of the half-life method arises when the tracer is a radioactive isotope with a *physical half-life* and is subject to a *biological half-life* (i.e., the time it takes for one-half the tracer to be removed through some biological process). Take, for example, the metabolism of iodine-131, which has a physical half-life of 8 days. There is also a biological half-life of iodine of about 4 days, which can also be considered first order. The problem posed, then, is what is the overall half-life of the tracer in the subject, i.e., *the effective half-life?* The reactions are as follows*:

$$^{131}\text{I} \xrightarrow{\ k_1\ } \text{products} \qquad (\text{physical decay}) t_{1/2} = 8 \text{ days}$$

$$^{131}\text{I} \xrightarrow{\ k_2\ } \text{products} \qquad (\text{biological decay}) t''_{1/2} = 4 \text{ days}$$

The rate of disappearance of ^{131}I is

(7.46)
$$-\frac{d[^{131}\text{I}]}{dt} = k_1[^{131}\text{I}] + k_2[^{131}\text{I}]$$

that is, the overall rate is a sum of the individual rates. Combining terms,

(7.47)
$$-\frac{d[^{131}\text{I}]}{dt} = [k_1 + k_2][^{131}\text{I}] = k[^{131}\text{I}]$$

where k is the overall rate constant, and the overall reaction is still first order. Therefore, T_{eff}, the overall half-life, is given by

(7.48)
$$T_{\text{eff}} = \frac{0.693}{k} = \frac{0.693}{k_1 + k_2}$$

Since both processes involved are first order, it is known that

(7.49)
$$k_1 = \frac{0.693}{T_p}$$

(7.50)
$$k_2 = \frac{0.693}{T_B}$$

---

*Here is another example of parallel, first-order reactions, this time using a different means of analysis.

Where T_p = physical half-life and T_B = biological half-life. From Eq. (7.48),

(7.51) $$k_1 + k_2 = \frac{0.693}{T_{\text{eff}}}$$

but

(7.52) $$k_1 + k_2 = \frac{0.693}{T_p} + \frac{0.693}{T_B}$$

by adding Equations 7.49 and 7.50. Therefore,

(7.53) $$\frac{0.693}{T_{\text{eff}}} = \frac{0.693}{T_p} + \frac{0.693}{T_B}$$

and finally, by dividing the equation by 0.693,

(7.54) $$\frac{1}{T_{\text{eff}}} = \frac{1}{T_p} + \frac{1}{T_B}$$

If $T_p = 8$ days and $T_B = 4$ days,

$$\frac{1}{T_{\text{eff}}} = \frac{1}{8} + \frac{1}{4}$$

or

$$T_{\text{eff}} = 2.67 \text{ days}$$

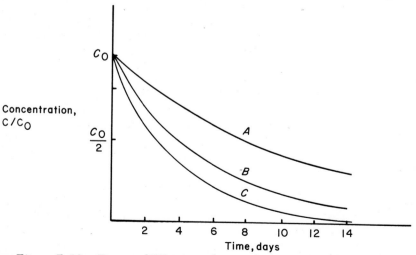

Figure 7–10. Decay of ^{131}I injected into a human subject. C shows overall decay; B, the biological decay (metabolism); A, the radioactive decay.

In 2.67 days, one-half of the tracer will be removed via both processes. Figure 7–10 shows the respective decays of the two processes and the composite decay.

7.1.9   Reversible Reactions

In theory all simple elementary reactions are reversible. For example, A reacts to form B:

(7.55) $$A \rightarrow B$$

If, however, we start with pure B, we find that B reacts to form A,

(7.56) $$B \rightarrow A$$

and in a mixture of A and B both reactions occur, or one can say that the reaction is reversible and so can be written with a double arrow:

$$A \rightleftharpoons B$$
$$B \rightleftharpoons A$$

If the reaction rate $A \rightarrow B$ is greater than that of reaction $B \rightarrow A$, then net B will be formed; that is, its concentration will increase with time while A decreases. As the concentration of A decreases, the rate of the forward reaction is decreased. As the concentration of B increases, the rate of the reverse reaction increases. A state of dynamic *equilibrium* is finally reached when the rate of the forward reaction is equal to the rate of the reverse. When *equilibrium* is reached, there will be no further net change in the reaction mixture; that is, the system possesses a macroscopic appearance of stability despite the fact that the reactants are changing to products and vice versa on the atomic or molecular level.

Let k_f equal the rate constant of the forward reaction in Equation (7.55) and k_r the rate constant of the reverse reaction. Then

(7.57) $$-\frac{dA}{dt} = k_f[A] - k_r[B]$$

that is, the net rate of disappearance of A is equal to its rate of disappearance by the forward reaction minus its rate of formation by the reverse reaction. If only A is present initially at a concentration A_0, then

$$B = A_0 - A$$

Therefore,

(7.58)
$$-\frac{dA}{dt} = k_f A - k_r(A_0 - A)$$
$$= (k_f + k_r)A - k_r A_0$$

Integration yields

(7.59)
$$-\int_{A_0}^{A} \frac{dA}{(k_f + k_r)A - k_r A_0} = \int_0^t dt$$

$$\frac{1}{k_f + k_r} \ln \left[(k_f + k_r)A - k_r A_0\right]\Big|_{A_0}^{A} = t$$

or

(7.60)
$$\ln \frac{k_f A_0}{(k_f + k_r)A - k_r A_0} = (k_f + k_r)t$$

At equilibrium, $dA/dt = 0$, so that

(7.61)
$$k_f A_{eq} = k_r B_{eq} = k_r(A_0 - A_{eq})$$

where the subscript eq refers to equilibrium concentrations. Then

(7.62)
$$A_{eq} = \frac{k_r}{k_f + k_r} A_0$$

Equation (7.60) may be rewritten by using this substitution to give

(7.63)
$$\ln \frac{A_0 - A_{eq}}{A - A_{eq}} = (k_f + k_r)t$$

The approach to equilibrium is a first-order process whose rate constant is the sum of the constants for the forward and reverse reactions. Consider again Equation (7.61):

$$k_f A_{eq} = k_r B_{eq}$$

Rearrangement yields

(7.64)
$$\frac{k_f}{k_r} = \frac{B_{eq}}{A_{eq}} = K$$

where K is the equilibrium constant, i.e., the equilibrium concentrations of products raised to the power of their coefficients in the overall reaction divided by the equilibrium concentrations of the reactants raised to the same appropriate power. For reactions in which both forward and reverse reactions take place to a measurable extent, knowledge of the equilibrium constant K and the rate constant for either the forward or the reverse reaction permits calculation of the other rate constant. If we know K and k_r, then

$$k_f = k_r K$$

It should be noted that although one knows the equilibrium expression of a reaction, it gives no information on the form of the rate law for either the forward or the reverse reaction except for simple elementary reactions.

There are many biological reactions which are reversible; for example, medical scientists are often measuring the rate that blood from the blood pool enters some organ and the rate that the blood returns to the pool. We may express this situation in compartments as follows:

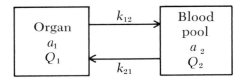

A tracer carried by the blood enters organ A. The rate constant for organ to blood is k_{12}* and blood to organ is k_{21}. The rate of tracer disappearance from organ A is

$$-\frac{da_1 Q_1}{dt} = k_{12} a_1 Q_1 - k_{21} a_2 Q_2$$

All calculations and expressions which are used for the chemical system apply to our compartmental system.

7.2   MULTICOMPARTMENTAL ANALYSIS

For most biological systems, one is dealing not with the simple situation that has been described but with the exchange of tracer among many compartments. Consider one of the most simple systems, the exchange of tracer between organ A and the blood pool and between the blood pool and the kidney, with the resulting clearance

---

*The notation used is that k_{ij} is the rate of transfer from compartment i to j.

of tracer. This scheme may be represented as

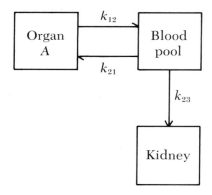

There is only one arrow from the blood pool to the kidney, as the tracer is subsequently removed from the system. The chemical analogy to this would be

$$A \underset{k_2}{\overset{k_1}{\rightleftharpoons}} B$$

$$B \underset{k_3}{\rightarrow} C$$

We must remember that if we consider only reaction $A \rightleftharpoons B$, A and B must reach their equilibrium concentrations consistent with the equilibrium constant. In this case, however, B is being removed from the system, so that more A reacts to form B; in fact, there is an equilibrium shift, and at infinite time all the activity is in the form of C. The rates of the above chemical reactions are:

(7.65)
$$-\frac{dA}{dt} = k_1[A] - k_2[B]$$

or

(7.66)
$$-\frac{dB}{dt} = k_1[A] - k_2[B] - k_3[B]$$

or

(7.67)
$$\frac{dC}{dt} = k_3[B]$$

Since B is subject to two reactions, it may be considered an intermediate whose concentration remains small; if either k_2 or k_3 is at least an order of magnitude greater than k_1 and is essentially constant, a steady-state approximation can be used, that is, $dB/dt = 0$. Therefore,

(7.68)
$$\frac{dB}{dt} = k_1[A] - k_2[B] - k_3[B] = 0$$

A value for B may be determined:

(7.69)
$$[B] = \frac{k_1[A]}{k_2 + k_3}$$

Substitution into the rate expressions for A and C gives

(7.70)
$$-\frac{dA}{dt} = \frac{dC}{dt} = \frac{k_3 k_1[A]}{k_2 + k_3}$$

There are some interesting possibilities that we may consider now. If $k_2 \gg k_3$, then

(7.71)
$$-\frac{dA}{dt} = \frac{k_1 k_3[A]}{k_2}$$

but $k_1/k_2 = K$. Therefore,

(7.72)
$$-\frac{dA}{dt} = k_3 K[A]$$

If $k_3 \gg k_2$, then

$$-\frac{dA}{dt} = \frac{k_1 k_3}{k_3}[A] = k_1[A]$$

which simply means that $A \rightarrow B$ is the slow or rate-determining step.

Let us apply our chemical results to our compartmental scheme. The experimental conditions may be as follows: (1) We may, by external detector, determine the rate of disappearance of tracer from organ A. (2) We may determine the rate of appearance of tracer in the urine. The rate expressions are

(7.73)
$$-\frac{da_1 Q_1}{dt} = k_{12} a_1 Q_1 - k_{21} a_2 Q_2$$

(7.74)
$$+\frac{da_3 Q_3}{dt} = k_{23} a_2 Q_2$$

and

(7.75)
$$\frac{da_2 Q_2}{dt} = k_{12} a_1 Q_1 - k_{21} a_2 Q_2 - k_{23} a_2 Q_2$$

Again, if the conditions are appropriate, we may apply a steady-state

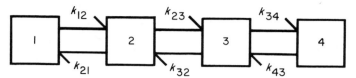

Figure 7–11. A catenary system of compartments.

approximation, or the equations may be integrated accurately (see Appendix B).

At this point, it should be evident that the compartmental concept offers many possibilities for describing many body systems in terms of compartmental reactions with direct analogy to chemical reactions. By constructing a compartmental scheme, a system may be described in terms of a set of differential (rate) equations.

Compartmental schemes have been divided into two general types—catenary and mammillary. In a *catenary system* the compartments are arranged in a chain, and tracer exchange occurs between adjacent compartments. Figure 7–11 illustrates a catenary system. A *mammillary system* is one in which a central compartment undergoes exchange reactions with surrounding compartments which do not exchange with each other. Figure 7–12 shows a typical mammillary scheme. Both these systems have chemical analogies. For a catenary system we have the following:

$$A \rightleftharpoons B$$
$$B \rightleftharpoons C$$
$$C \rightleftharpoons D$$
$$D \rightleftharpoons E$$

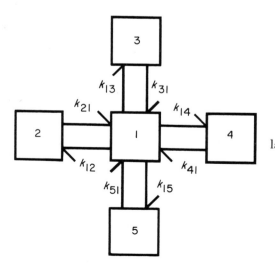

Figure 7–12. A mammillary system of compartments.

or simply $A \rightleftharpoons B \rightleftharpoons C \rightleftharpoons D \rightleftharpoons E$, that is, a set of reactions such that the product of one reaction is the reactant for the next. Such a set is, of course, series or consecutive reactions, which have been discussed previously.

A mammillary system may be represented by the following chemical scheme of parallel reactions:

$$A \rightleftharpoons B$$
$$A \rightleftharpoons C$$
$$A \rightleftharpoons D$$
$$A \rightleftharpoons E$$

or

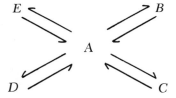

Although these two systems have been distinguished as two separate situations, mathematically there is no difference. The treatment is exactly the same and is given in Appendix B, where solutions to the general kinetic model of series and parallel reactions are discussed. It should be noted that both compartmental analysis and kinetic analysis of this type depend on all reactions being first order and are not applicable to a combination of first-order and zero-order reactions. The general kinetic scheme of first-order series and parallel reactions is

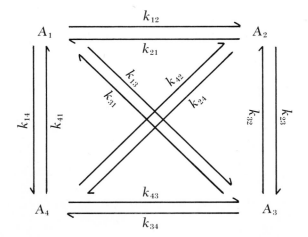

where A_1, A_2, . . . , A_j are the concentrations of m substances and k_{ij} is the first-order rate constant for the reaction of A_i to A_j.

The general solution to this scheme is obviously the solution to a set of m rate equations of the form

$$\frac{dA_1}{dt} = k_{21}[A_2] + k_{31}[A_3] + k_{41}[A_4] + \cdots$$
$$- k_{12}[A_1] - k_{13}[A_1] - k_{14}[A_1] + \cdots$$

for the rate of formation of A. If all reactions do not occur in the system of interest, these rate constants are simply zero.

Consider again a catenary system, which in terms of our general model may be represented as

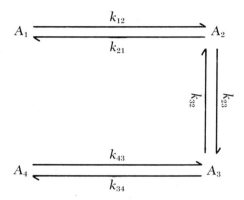

where k_{13}, k_{31}, k_{42}, k_{24}, k_{14}, and k_{41} are all zero. For a mammillary system, represented

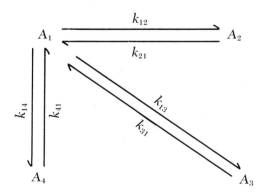

where k_{42}, k_{24}, k_{23}, k_{32}, k_{43}, and k_{34} are zero. Let it suffice at this point to say that although the terminology for these systems is indeed picturesque, the mathematical treatment is exactly the same. In addition, the general model incorporates systems that are both catenary and mammillary.

7.2.1 Mechanisms and Models

Kinetics provides the most general method for determining the mechanism of a reaction. The mechanism, as we have defined it previously, is the set of elementary reactions that taken together detail the pathway from reactant to product. Determination of the rate law, by studying the rate of a reaction as a function of concentrations, temperature, and any other operating variable, gives the species involved in the rate-determining step(s). There are, however, several ways of interpreting kinetic data, and frequently several mechanisms may be consistent with the data. Suppose we have studied the stoichiometric reaction

$$A + 2B \rightarrow C$$

and we find that the rate is

$$-\frac{dA}{dt} = k[A][B]$$

that is, the rate is proportional to the concentration of A and B. The kinetics of this reaction can be explained by the following steps:

$$A + B \rightarrow AB \quad \text{slow}$$
$$AB + B \rightarrow C \quad \text{rapid}$$

The slow step determines the rate law and the order of the reaction. An equally consistent mechanism, however, is

$$A + B \rightarrow D \quad \text{slow}$$
$$A + D \rightarrow E \quad \text{rapid}$$
$$E + B \rightarrow C + A \quad \text{rapid}$$

Kinetics does not give an unambiguous answer concerning individual steps. The mechanism is a mental model devised to fit the facts. Decisions as to which mechanism to accept often involve further experimentation in which one attempts to vary the conditions in such a way that the proposed intermediates may be detected. Like any theory, a mechanism currently in vogue may be discounted by new experimental evidence.

Model building for a biological system is subject to the same rules as mechanisms of a chemical reaction. Many models may be proposed for the same experimental data. The rate expression, together with imagination, general medical and biological experience, and principles of the tracer method, is used to deduce a model for a system of interest.

Suppose we have a presumed multicompartmental system. We begin by the introduction of tracer material in one compartment. At suitable time intervals, samples may be withdrawn from the initial compartment and other presumed compartments and analyzed. One may then infer from the results the net transport of tracer between compartments. But in practice, a somewhat different approach has been preferred. The experimental points are fitted to a smooth curve whose shape is predicted by the assumed kinetic model. This procedure involves the solution of differential equations involving specific activity for the compartments as a function of time. Again this is precisely the chemical analogy. Once we have the experimental curves, the problem becomes one of devising a model that predicts those curves.

There are a few simple examples that are considered here. Suppose there is a system of three compartments

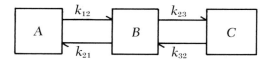

Let us propose different experimental conditions and results and the probable kinetic model.

Case I

Tracer material is injected into compartment A, and its net disappearance is observed. In addition, tracer concentration is observed in compartments B and C. Figure 7–13 shows the curves of tracer concentration versus time obtained. The specific activity in compartment A is a simple exponential decay; specific activity in compartment B passes through a maximum; and specific activity in compartment C is an exponentially increasing curve preceded by an induction period, that is, a very slow initial rate of formation. These results would fit the model

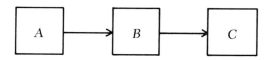

that is, the case where $k_{21} = k_{32} = 0$. The rate equations and solutions are given in Sec. 7.1.7.

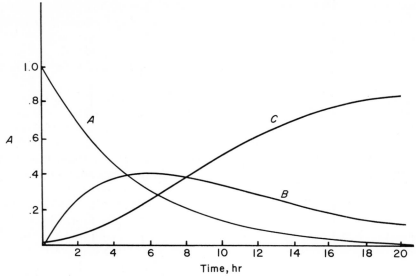

Figure 7–13. Tracer concentration in compartments A, B, and C when initially injected into A.

Case II

Tracer material is again injected into compartment A, and the specific activity or tracer concentration is determined in all three compartments. The specific activity of A decays exponentially. The specific activity of B decays as some composite exponential; the specific activity of C increases exponentially (Refer to Figure 7–8). There are two similar mechanisms that we may propose:

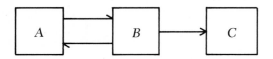

with $k_{32} = 0$. If $k_{21} \gg k_{23}$, then an equilibrium exists between A and B, and $B \rightarrow C$ is the rate-determining step. The exact rate expressions are given in Section 7.1.

Case III

Tracer material is injected into compartment B, and its net disappearance is observed. The concentration of C increases exponentially; disappearance of tracer in A and B appears to be a complicated exponential function. Further, a plot of log B versus time gives a curve, but with increasing time appears to be linear. The fact that the semilog plot appears to be linear after some time t suggests a parallel first-order model. After a time t, the formation of C is dependent upon only one reaction. Therefore, this straight-line portion is extrapolated back to $t = 0$. The extrapolated values are subtracted from the experimental curve and the values plotted to give the formation of C by another reaction. The mechanism is proposed as

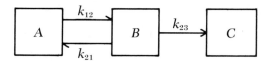

where $k_{32} = 0$.

Suppose only the disappearance of tracer in compartment B could be observed. It would be impossible to choose between

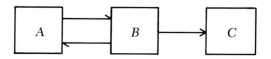

and

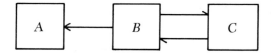

The exact form of the rate equation is given in Section 7.1.

The general solution for the proposed three-compartmental system, where no k's are assumed zero, is given in Appendix B. We must also note that a pulse input has been imposed. This, of course, is not a general case. The general equation for the net disappearance of a material in a series of compartments is

$$(7.76) \qquad -\frac{dS}{dt} = I + \Sigma k_{ij} S_i - \Sigma k_{ji} S_j$$

where S is any substance, I is the input function, and $\Sigma k_{ji} S_j$ the sum of all rates of appearance. For pulse input, $I = 0$. The whole process of compartmental analysis involves determination of I and all k_{ij} and S_i or, at least, the interrelationship of all of them from experimental quantities. An example of such is the clearance of technetium from the body (also see Sec. 7.1.2.).

BIBLIOGRAPHY

The bibliography for Chapter 7 appears at the end of Chapter 8.

Chapter 8

QUANTITATIVE EXAMPLES OF BIOMEDICAL KINETICS

8.1 EXAMPLES OF MULTICOMPARTMENTAL MODELS FROM KINETIC DATA

The greatest potential of kinetic data is the extraction by inference, deduction, and imagination of a model or mechanism that will describe the system and predict the observed values. When large numbers of components are involved, the quantitative aspects are more easily handled by a computer program. Since most reactions between compartments are assumed first order, an experimentally determined rate curve that is a composite of series and parallel first-order reactions is usually fitted to a program utilizing sums of exponentials. For systems involving only a few compartments, manual solutions are possible, as we have already indicated.

8.1.1 Iodine Kinetics

There are some important kinetic studies that have resulted in models for the systems under study. First we shall consider iodine kinetic studies, which have historic importance, being some of the first processes subjected to extensive tracer investigations. The radio-isotope ^{125}I in the form of iodide I^- was injected intravenously and blood samples were analyzed to determine the amount of tracer in plasma. The thyroid was externally monitored with a counter. The appearance of tracer in urine was also observed. The general model is a three-compartment system.

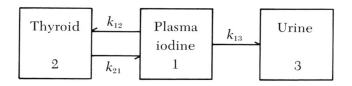

The thyroid compartment is larger than the two other compartments — actually the thyroid is more complicated and is probably best represented by three compartments itself. This simplified model is analogous to the following reactions:

$$A \underset{k_{12}}{\overset{k_{21}}{\rightleftharpoons}} B$$

$$B \xrightarrow{k_{13}} C$$

where A is thyroid, B plasma, and C urine. The rate of disappearance of tracer from plasma is

$$-\frac{dB}{dt} = k_{13}[B] + k_{12}[B] - k_{21}[A]$$

and from the thyroid A is

$$-\frac{dA}{dt} = k_{21}[B] - k_{12}[A]$$

and the rate of appearance into the urine C is

$$\frac{dC}{dt} = k_{13}[B]$$

Again we have a set of coupled differential equations. When solved simultaneously, they should predict the experimental curves if the model is correct.

8.1.2   Technetium Kinetics

Another example is the understanding of the behavior of technetium 99m-pertechnetate $[^{99m}TcO_4^-]$ in humans. It has been assumed that $^{99m}TcO_4^-$, which is a popular radiopharmaceutical used for brain-tumor localization, behaves according to the following simple model:

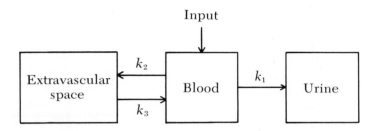

This extravascular space includes a nonspecific distribution throughout the interstitial space and a specific accretion of $^{99m}TcO_4^-$ by certain organs, namely, thyroid, stomach, salivary glands, and choroid plexus (the latter may be difficult to distinguish from a brain tumor). Thus, patients are pretreated with iodine or perchlorate (ions of similar size and charge to pertechnetate) to saturate these organs and thereby block their uptake.

If the blocking is complete, the model can be reduced to

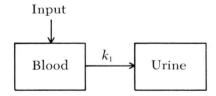

In this case, the model reduces to a simple first-order transfer from the blood compartments to the urine compartment. Thus, the simple model for the patient pretreated with iodide or perchlorate suggests a rate law

$$\ln \frac{(\text{Blood activity})_{t=0}}{(\text{Blood activity})_{\text{time } t}} = k_1 t$$

Therefore, by plotting $\ln B_0/B$ against t, one can (1) see whether the simple model holds for the pretreated case; (2) calculate k_{-1}, if the simple model is valid.

Plots of $\ln B_0/B$ against t for both the pretreated and the non-pretreated patients are shown in Figure 8–1. The curve for the pre-

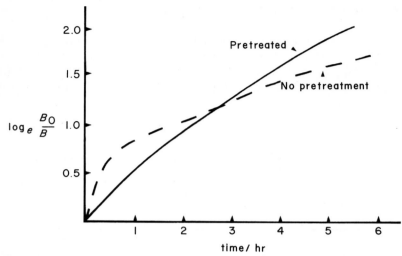

Figure 8–1. A plot of $\ln B_0/B$ versus time.

treated patients approaches linearity, so in this case the simple model is valid with $k_1 \approx 0.2 \text{ hr}^{-1}$; the slight variation from linearity shows that there is a fraction of the extravascular space that is not blocked (e.g., lymphatics, interstitial fluid, etc.).

Initially there is no activity in the extravascular space, so when the time is small, the model approximates to

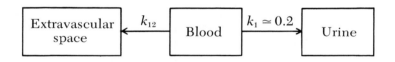

Therefore, in this case, when t is small:

$$\ln \frac{B_0}{B} = (0.2 + k_2)\, t$$

that is,

$$k_2 \simeq 1.45 \text{ hr}^{-1}$$

8.1.3 Ferrokinetics

Iron-59 binds to transferrin (a plasma protein) and is used as the tracer. Organ uptake is studied by external monitoring, whereas the behavior of the tracer in the plasma can be evaluated by removing blood samples and counting them in a well counter. Studies were performed to identify the exact form of the general model.

Erythropoiesis

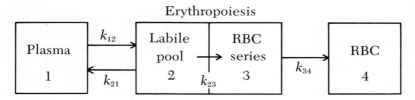

As the exchange from 3 to 4 is assumed to be very fast, this model of plasma ion disappearance is a simple four-compartment model.

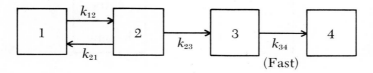

Rate of egress of radioiron from plasma should be a simple sum of two exponential decays. In normal patients the simple model is correct, and the egress is governed by two decays with a half-life of $\simeq 8$ hr and 4 days. In hemolytic disease, however, this simple model is not consistent with the observed events; i.e., the egress is not explained by two exponentials as the plasma activity reaches an equilibrium value. In this case the following model is more consistent with the observed data, when the hemolysis of circulating erythrocytes results

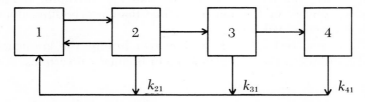

in feedback of the radioiron, and this feedback (from compartments 2, 3, and 4 to 1) results in the final equilibrium value rather than the egress to zero concentration.

8.1.4   Bilirubin Kinetics

A fourth example of a proposed model for a system of interest is the study of bilirubin turnover. Bilirubin is labeled with ^{14}C and injected as albumin-bilirubin. Plasma tracer activity was determined by removal of blood samples, and the rate of disappearance of plasma bilirubin was determined. This rate curve was a composite exponential decay which could be resolved into two components. The following two models were proposed:

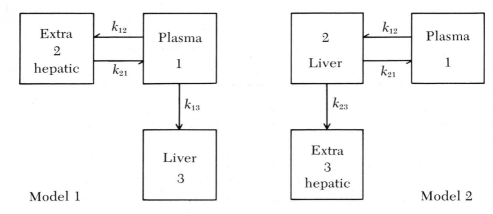

Both are consistent with the data. Other evidence, however, has suggested that Model 2 represents the actual physiologic situation. For example, equilibrium studies of bilirubin-^{3}H distribution found that 87 per cent of injected bilirubin was recovered from hepatic tissue, indicating that little label entered nonhepatic tissue.

Comparison of patients with Gilbert's disease (idiopathic hyperbilirubinemia) with normal patients showed that in the diseased state k_{21} is increased six fold in the diseased patients, whereas k_{12} and k_{23} remain essentially the same. These data suggest that in Gilbert's disease the liver is unable to hold the bilirubin, and so the bilirubin returns to the plasma rather than proceeding along the normal excretory pathway.

These few examples show, however briefly, the application of model building to the understanding of biomedical systems.

8.2 ALTERNATIVE APPROACHES TO ANALYSIS

In the preceding section, we have stressed compartmental analysis and models. An attempt was made to delineate the transport of material from one functional site to another. There is, however, another approach to analysis and use of kinetic data which has received some attention. Noncompartmental analysis involves determination of the input function and subsequent output function. In many respects, this approach is similar to the black box problems of the physical sciences. Figure 8–2 shows an input function and subsequent output function. The shape of the output and its change from input is obviously a function of the system as a whole. A specific

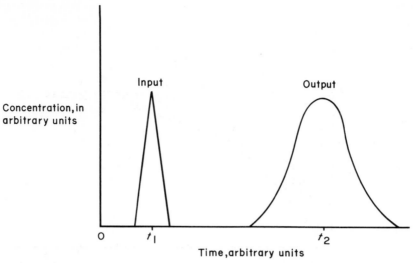

Figure 8–2. A complex output function derived from a pulse input function.

example of this approach will be given in the following section, where the measurement of regional blood flow is given to illustrate the different approaches to interpretation of kinetic data. We shall develop various approaches to show the progress in the mathematical manipulation of the experimental data as the data are refined.

8.2.1   Measurement of Regional Cerebral Blood Flow

The general methods of measuring regional blood flow were discussed in the introduction to this chapter. It was pointed out that intuition alone is not always adequate to evaluate complex kinetic data. Thereby the need was demonstrated for more rigorous treatment of the data. Now that we have discussed some principles for handling such complex data, we shall use cerebral blood flow to illustrate proper application of these tracer principles.

For the measurement of regional cerebral blood flow nitrous oxide (N_2O) in low concentration was continuously inhaled for a period of approximately 10 min. During the inhalation time, samples of arterial and internal jugular venous blood were withdrawn and the concentration of the nitrous oxide measured by conventional chemical means.

From the experimental curves, which are of the type illustrated in Figure 8–3, the average cerebral blood flow can be measured by the following analysis. Let

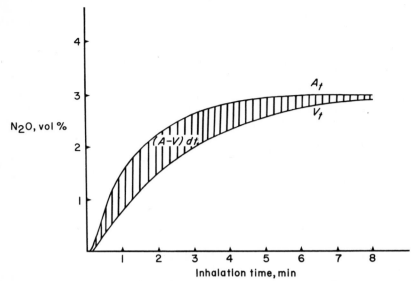

Figure 8–3. Arterial and venous concentrations of nitrous oxide measured in a patient inhaling a constant percentage of nitrous oxide.

(Q_B) = quantity of N_2O taken up by brain in time t measured from
$$t = 0$$
(Q_A) = quantity of N_2O brought to brain by arterial blood in time t
(Q_V) = quantity of N_2O carried away from brain by venous blood in time t
A_t = arterial N_2O concentration at time t
V_t = venous N_2O concentration at time t
W = weight of brain
TF = total cerebral blood flow (cc/min)
CBF = cerebral blood flow (cc/100 gm brain tissue/min)

From the Fick principle of material conservation

(8.1)
$$(Q_B) = (Q_A) - (Q_V)$$

Now A_t and V_t are variables, where $\int_0^\infty V_t$ is the function of the tracer leaving the brain. Therefore,

(8.2)
$$(Q_A) = TF \int_0^t A_t \, dt$$

(8.3)
$$(Q_V) = TF \int_0^t V_t \, dt$$

Therefore,

$$(8.4) \qquad (Q_B) = TF \int_0^t (A_t - V_t)\, dt$$

$$CBF = \frac{TF}{W}$$

$$(8.5) \qquad CBF = \frac{(Q_B)/W}{\displaystyle\int_0^t (A_t - V_t)\, dt}$$

The integral in the denominator is simply the shaded area in Figure 8–3, and if the time t is sufficient for equilibrium to have been reached

$$(8.6) \qquad \frac{(Q_B)}{W} = V_t s$$

where s is the partition function for nitrous oxide between brain and blood. The partition function is simply

$$s = \frac{\text{concentration of nitrous oxide in brain at equilibrium}}{\text{concentration of nitrous oxide in blood at equilibrium}}$$

$$(8.7) \qquad = \frac{Q_B/W}{V_t}$$

Therefore,

$$(8.8) \qquad CBF = \frac{100\, V_\infty s}{\text{area between curves}} \qquad \text{cc}/(100 \text{ gm of brain})(\text{min})$$

where V_∞ is the concentration of tracer in venous blood at equilibrium. This expression for the determination of cerebral blood flow is derived without any assumptions as to the nature of the input function. The other assumption is that there is no recirculation of N_2O.

Using this method, one can only obtain a measure for total cerebral blood flow, as discussed in the introduction to this chapter; in order to determine regional cerebral blood flow, it is necessary to use an inert radioactive gas and observe the uptake and egress of the indicator from various regions of the brain with collimated radioactivity detectors. It is probably not possible to determine regional cerebral blood flow rCBF from the inhalation technique since a considerable portion of the detected radioactivity will be emanating from noncerebral tissue. This may be overcome by injecting a bolus of the tracer (dissolved in physiological saline) directly into the internal carotid artery. In this case, the curve obtained for activity in a particular volume plotted against time after injection looks like Figure 6–6. (p. 117)

Application of the Fick principle gives the relation

(8.9)
$$\frac{d(Q_B)_t}{dt} = TF\,(A_t - V_t)$$

A_t in this case is always equal to zero, as the pulse of activity in the arterial blood arrives at the brain at $t = 0$ and there is negligible recirculation of the activity as the inert gas is ventilated by the lung. Now,

(8.10)
$$TF = \frac{rCBF \cdot W}{100}$$

where rCBF is the regional cerebral blood flow in cubic centimeters per 100 gm of brain per minute, or

(8.11) $\quad rCBF = 100\,\dfrac{[d(Q_B)_t/dt]}{V_t W}$ ml/(100 gm) (min)

To continue the analysis, an assumption has to be made regarding the partition function. In our calculation of the theory for the inhalation method we stated that at equilibrium

$$s = \frac{Q_B}{WV_t}$$

If we assume that diffusion between the brain and blood is so fast that equilibrium is always in effect, we can substitute $V_t W = Q_B/s$ into our expression for rCBF. Therefore,

(8.12) $\quad rCBF = \dfrac{100\,s\,[d(Q_B)_t/dt]}{(Q_B)_t}\qquad$ ml/(100 gm)(min)

as $d(Q_B)_t/dt$ is simply the tangent to the clearance curve at time t, whereas $(Q_B)_t$ is simply the observed counting rate at time t. The rCBF can, therefore, be calculated. For example, using the curve shown in Figure 8–1, at $t = 1$ min,

$$\frac{(dQ_B)_t}{dt} = \frac{1100}{2.75} = 400$$

whereas $(Q_B)_t = 700$;

$$\text{therefore, } rCBF = 100\,s\,\frac{-400}{700} = 57\,s \qquad \text{ml/(100 gm)(min)}$$

However, at $t = 6$ min,

$$\frac{(dQ_B)_t}{dt} = \frac{330}{12.5} = 26.4$$

whereas $(Q_B)_t = 180$; therefore,

$$rCBF = \frac{100s \cdot 26.4}{180} = 14.7s \qquad ml/(100\ gm)(min)$$

It is seen that by calculating the flow rate at two different times, one obtains two different flow rates.

It is found that the brain does not behave as a single compartment. A simple analysis (of the type performed in Figure 8–4) of the clearance curve of the inert gas shows that the curve is the sum of two mono-exponential components, and this leads to a two-compartment model of the inert-gas washout, i.e.,

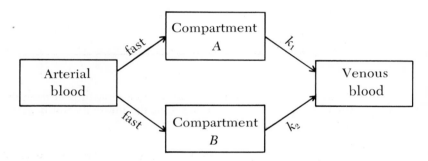

The tracer is washed out of these compartments with different rate constants and with different flow rates. An estimate of the relative sizes of these two compartments can be seen by plotting the washout curve on semilogarithmic paper, drawing the tangent to the curve at large times to give one of the components — the slow component (see Sec. 7.1.7). The other (fast) component can be found by subtraction of the slow component from the experimentally determined curve. A typical analysis of this type is shown in Figure 8–4. Although this "curve stripping" technique is adequate for a crude analysis, it tends to introduce a systematic error by giving undue emphasis on the slow compartment — where at long times the low count rate (poor statistics) results in less accurate measurements. Values of flow from each of the two compartments can be obtained by analyzing each component of the washout curve separately.

It has been postulated that the two compartments represent the white matter and gray matter of the brain, and that the partition co-efficients should differ for each of the compartments, since the solu-

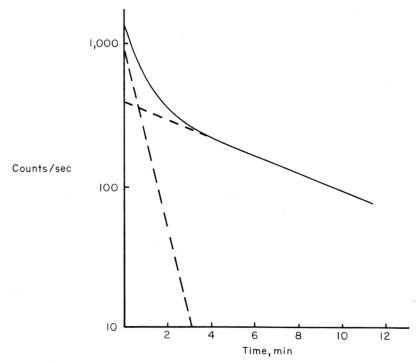

Figure 8–4. A typical curve for the washout of an inert tracer from the brain. The tracer was injected as a pulse input via the internal carotid artery. Curve stripping of a typical inert washout curve.

bility of inert gases differs from white to gray matter. One way of obtaining values of rCBF is, therefore, to obtain two flow rates for each region and assume (1) that the two compartments have different partition coefficients; (2) that these compartments are really gray and white matter, and the partition coefficient for gray and white matter can be applied.

In this method we have (1) performed a compartmental analysis and (2) assumed that the model is correct. There is, however, a second method to analyze the data that makes no assumptions as to the compartments in the brain and simply considers the transit of the indicator through the brain. The mathematics of this method, which gives a value of mean flow per unit volume, is given below.

If q_0 is the amount of tracer injected, then the amount of tracer leaving the volume of interest in a time, dt, is $q_0 h(t)\, dt$, where $h(t)$ is the frequency function of the tracer (and we assume also of the thing being traced). As this amount must equal the product of blood flow out of the system $F\, dt$ and concentration V_t, then

$$q_0 h(t)\, dt = FV_t\, dt$$

(8.13)
$$h(t) = \frac{F}{q_0} V_t$$

By integration,

(8.14)
$$q_0 \int_0^t h(t) \, dt = F \int_0^t V_t \, dt$$

and we define

(8.15)
$$H(t) = \int_0^t h(t) \, dt$$

The amount of tracer remaining in the brain at time t is given by

(8.16)
$$(Q_B)_t = q_0 [1 - H(t)]$$

Therefore,

(8.17)
$$\int_0^\infty (Q_B)_t \, dt = q_0 \int_0^\infty [1 - H(t)] \, dt$$

By integration we can show that

(8.18)
$$\int_0^\infty [1 - H(t)] \, dt = \bar{t}$$

where $\bar{t}$ is the mean transit time of the tracer through the system and is equal to V/F, the volume of the system divided by the flow rate through the system.

Therefore,

(8.19)
$$\frac{F}{V} = \frac{q_0}{\displaystyle\int_0^\infty (Q_B)_t \, dt}$$

As the tracer is distributed between volume V and the tissue volume V_i, $s = V/V_i$ and so

(8.20)
$$\frac{F}{sV_i} = \frac{q_0}{\displaystyle\int_0^\infty (Q_B)_t \, dt} = \frac{\text{peak height at zero time}}{\text{area under curve}}$$

This analysis, which considers the volume as a whole without any regard to compartments, obtains a mean value for the flow per unit volume of brain. In deriving this expression, however, assumptions are also made: (1) the tracer behaves as the thing being traced; (2) $s = V/V_i$.

This latter method, in which one considers solely the transit function, can be referred to as the "lumping" process, while the former method of dividing the system in compartments can be referred to as "splitting."

This review of the theory of blood flow has been given to demonstrate methods to manipulate data in tracer kinetics and to show that assumptions are necessary to analyze the data by either the lumping or the splitting method. It is imperative to be alert to all the assumptions in any kinetic analysis involving tracers. Frequently, small manipulations of either the tracer or the data result in erroneous conclusions since they no longer obey the initial assumptions. The development of blood-flow measurement theory to its present stage emphasizes many of the points discussed in the preceding parts of this book.

BIBLIOGRAPHY FOR CHAPTERS 6, 7, and 8.

Barrett, P. V. D., P. D. Berk, M. Menken, and N. I. Berlin: Bilirubin Turnover Studies in Normal and Pathologic States Using Bilirubin ^{14}C, *Ann. Intern. Med.*, **68**:355 (1968).

Berlin, N. I.: Life Span of the Red Cell, in C. Bishop and B. M. Surgenor (eds.), "The Red Blood Cell," 423–450, Academic Press, Inc., New York, 1964.

Bing, R. J., A. Bemmish, G. Bluemchen, A. Cohen, J. P. Gallagher, and E. J. Zaleski: The Determination of Coronary Flow Equivalent with Coincidence Counting Technique, *Circulation*, **24**:833 (1964).

Blaufox, M. D., A. L. Orvis, and C. A. Owen, Jr.: Compartment Analysis of Radiorenogram and Distribution of Hippuran I-131 in Dogs, *Amer. J. Physiol.*, **204**:1059 (1963).

Brownell, G. L., M. Berman and J. S. Robertson: Nomenclature for Tracer Kinetics, *Int. J. Appl. Radiat. Isotop.*, **19**:249 (1968).

Fick, A.: Ueber die messung des Blut quantums in den Herzventrikeln. Sitzungsb der phys-med. ges zu. Wurzburg 1870.

Finch, C. A. et al.: The Ferrokinetic Approach to Anemia, in "Series Hematologic, Number 6," The Williams & Wilkins Co., Baltimore, 1965.

Fox, I. J., and E. H. Wood: Application of Dilution Curves Recorded from the Right Side of the Heart or Venous Circulation with the Aid of a New Indicator Dye, *Proc. Staff Meet. Mayo Clinic*, **32**:541 (1957).

Frost, A. A., and R. G. Pearson: "Kinetics and Mechanism," John Wiley & Sons, Inc., New York, 1961.

Hamilton, W. F., and J. W. Remington: Comparison of the Time Concentration Curves in Arterial Blood of Diffusible and Non-diffusible Substances When Injected at a Constant Rate and When Injected Instantaneously, *Amer. J. Physiol.*, **148**:35 (1947).

———— et al.: Comparison of Fick and Dye Injection Method of Measuring Cardiac Output in Man, *Amer. J. Physiol.*, **153**:309 (1948).

———— J. W. Moore, J. M. Kinsman, and R. G. Spurling: Studies on the Circulation. IV. Further Analysis of the Injection Method, and of Changes in Hemodynamics under Physiological and Pathological Conditions, *Amer. J. Physiol.*, **99**:534 (1932).

Hill, R., J. Clifton, T. Gallager, and E. J. Potchen: Regional Cerebral Blood Flow in Man. II. Data Acquisition and Analysis, *Arch. Neurol.*, **20**:384 (1969).

Hower, H. W.: Renal Clearances of Substituted Hippuric Acid Derivatives and Other Aromatic Acids in Dog and Man, *J. Clin. Invest.*, **24**:388 (1965).

Kety, S. S., and C. F. Schmidt: Determination of Cerebral Blood Flow in Man by the Use of Nitrous Oxide in Low Concentrations, *Amer. J. Physiol.*, **143**:53 (1945).

Lowry, T. M., and W. T. John: Studies of Dynamic Isomerism. Part XII. The Equations for Two Consecutive Unimolecular Changes, *J. Chem. Soc. (London)*, **97**:2634 (1910).

Lundquist, F., and H. Woethers: The Kinetics of Alcohol Elimination in Man, *Acta Pharmacol. Toxicol.*, **14**:265 (1958).

Matsen, F. A., and J. L. Franklin: A General Theory of Coupled Sets of First-order Reactions, *J Amer. Chem. Soc.*, **72**:3337 (1950).

Moore, F. D.: "The Body Cell Mass and Its Supporting Environment," W. B. Saunders Company, Philadelphia, 1963.

Penner, J. A.: Investigation of Erythrocyte Turnover with Selenium-75 Labeled Methionine, *J. Lab. Clin. Med.*, **67**:427 (1966).

Potchen, E. J., D. O. Davis, M. H. Adatepe, and J. Taveras: Shunt Flow in Glioblastoma — A Study in Regional Autoregulation, *Invest. Radiol.*, **3**:186 (1969).

_____, _____, T. Wharton, R. Hill, and J. Taveras: Regional Cerebral Blood Flow in Man. I. A Study of Xenon-133 Method, *Arch. Neurol.*, **20**:378 (1969).

Rescigno, A., and G. Andsegre: "Drug and Tracer Kinetics," Blaisdell Publishing Company, a division of Ginn and Company, Waltham, Mass. 1966.

Saperstein, L. A., and L. E. Moses: Cerebral and Cephalic Blood Flow in Man: Basic Considerations of the Indicator Fractionation Technique in Dynamic Clinical Studies with Isotopes, Proceedings of symposium held at Oak Ridge Institute of Nuclear Studies, October 21–25, 1963.

Sheppard, C. W.: "Basic Principles of the Tracer Method," John Wiley & Sons, Inc., New York, 1962.

Small, W. J., and M. C. Verloop: Determination of the Blood Volume Using Radioactive ^{51}Cr, *J. Lab. Clin. Med.*, **46**:255 (1956).

Sterling, K., and S. J. Gray: Determination of the Circulating Red Cell Volume or Mass by Radioactive Chromium, *J. Clin. Invest.*, **29**:1614 (1950).

Stewart, G. N.: Researches on the Circulation Time and on the Influences Which Affect It. IV. The Output of the Heart, *J. Physiol.*, **22**:159 (1897).

_____: Pulmonary Circulation Time: Quantity of Blood in the Lungs and the Output of the Heart, *Amer. J. Physiol.*, **58**:20 (1921).

Taplin, G. V., D. E. Johnson, E. K. Dow, and H. S. Keplon: Suspension of Radioalbumin Aggregates for Photoscanning the Liver, Spleen, Lung and Other Organs, *J. Nucl. Med.*, **5**:259 (1964).

Wagner, H. N., Jr., E. L. Jones, D. E. Tow, and J. K. Langon: Preliminary Report: A Method for Study of Peripheral Circulation in Man, *J. Nucl. Med.*, **6**:150 (1965).

Welch, M. J., M. H. Adatepe, and E. J. Potchen: A Clinical Analysis of Technetium (^{99m}TcO$_4^-$) Kinetics: The Effect of Heavy Ion Pretreatment, *J. Appl. Radiat. Isot.*, **20**:437 (1969).

Zierler, K. L.: Circulation Times and the Theory of Indicator Dilution Methods for Determining Blood Flow and Volume, in "Handbook of Physiology Circulation," vol. I, p. 585, American Physiological Society, Washington, 1962.

Appendix A

Critical Values of *t* for Tests of Significance

Degrees of Freedom	Probability Levels		
	90% $t_{0.10}$	95% $t_{0.05}$	99% $t_{0.01}$
1	6.31	12.71	63.66
2	2.92	4.30	9.93
3	2.35	3.18	5.84
4	2.13	2.78	4.60
5	2.02	2.57	4.03
6	1.94	2.45	3.71
7	1.90	2.37	3.50
8	1.86	2.31	3.36
9	1.83	2.26	3.25
10	1.81	2.23	3.17
15	1.75	2.13	2.95
20	1.73	2.09	2.85
30	1.70	2.04	2.75
40	1.68	2.02	2.70
60	1.67	2.00	2.66
∞	1.65	1.96	2.58

TABLE A–2.

Simplified Confidence Limits Table°

n	Degrees of Freedom $n-1$	Confidence Levels		
		90% f_{90}	95% f_{95}	99% f_{99}
2	1	4.48	9.01	45.15
3	2	1.69	2.49	5.74
4	3	1.18	1.59	2.92
5	4	0.951	1.24	2.05
6	5	0.825	1.04	1.64
7	6	0.732	0.925	1.40
8	7	0.671	0.837	1.24
9	8	0.620	0.770	1.12
10	9	0.579	0.715	1.03
15	14	0.455	0.556	0.762
20	19	0.378	0.456	0.622
40	39	0.305	0.366	0.494
60	59	0.214	0.256	0.341

°With n individual values $[(n-1)$ degrees of freedom] for calculating sample mean $\bar{x}$, the true mean may be expected to lie within the range $\bar{x} \neq fs$ with a % probability as indicated by the f subscript. f's are simply t/rn using appropriate values of t.

TABLE A–3.
Values of χ^2

Degrees of Freedom	Probability Levels		
	90% $\chi^2_{0.10}$	95% $\chi^2_{0.05}$	99% $\chi^2_{0.01}$
1	2.71	3.84	6.64
2	4.61	5.99	9.21
3	6.25	7.82	11.34
4	7.78	9.49	13.28
5	9.24	11.07	15.09
6	10.65	12.59	16.81
7	12.02	14.07	18.48
8	13.36	15.51	20.09
9	14.68	16.92	21.67
10	15.99	18.31	23.21
15	22.31	25.00	30.58
20	28.41	31.41	37.57
25	34.38	37.65	44.31
30	40.26	43.77	50.89

Appendix *B*

THE GENERAL SOLUTION FOR A MULTICOMPARTMENTAL FIRST-ORDER SYSTEM

As discussed in Section 6.4, the most general form of a multicompartmental model is one where the tracer in every compartment can exchange with every other compartment. From this general model, particular models can be solved by putting various rate constants equal to zero.

For example, for a four-compartment system

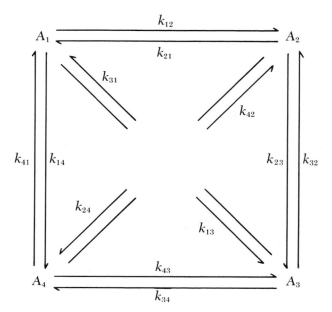

The rate equation for the first compartment is

(B.1) $\quad -\dfrac{dA_1}{dt} = -k_{21}A_2 - k_{31}A_3 - k_{41}A_4 + k_{12}A_1 + k_{13}A_1 + k_{14}A_1$

If the model consists of m compartments, then the rate equation from any compartment is

(B.2) $\quad \dfrac{dA_i}{dt} = (-k_{i1} - k_{i2} - k_{i3} - \cdots - k_{im}) A_i + k_{1i}A_1 + k_{2i}A_2 + k_{3i}A_3 +$

$$\cdots + k_{mi}A_m$$

Equation (B.2) can be written

(B.3) $\quad \dfrac{dA_i}{dt} = - \sum_{j=1}^{m} K_{ij}A_j \qquad i = 1, 2, 3, \ldots, m$

where $K_{ij} = -k_{ij}$ and $K_{ii} = \sum_p k_{ip}$ $(p = 1, 2, 3, \ldots, m)$. Equations of this type can be integrated and the integral shown to be

(B.4) $\quad A_i = \sum_{r=1}^{m} B_{ir}e^{-\lambda_r t}$

The solution A_i can be considered to be a vector made up of m components in concentration space.

If we assume there is one solution of Equation (B.2) of the type

(B.5) $\quad A_i = B_i e^{-\lambda t}$

and if this solution is substituted into Equation (B.3), we have a series of equations

(B.6) $\quad -\lambda B_i + \sum_{j=1}^{m} K_{ij} B_j = 0$

or

(B.7) $\quad \sum_{j=1}^{m} (K_{ij} - \delta_{ij} \lambda) B_j = 0 \qquad i = 1, 2, 3, \ldots, m$

where $\delta_{ij} = 1$ if $i = j$, or 0 if $i \neq j$.

If Equation (B.7) can be solved, the following condition must be satisfied:

(B.8) $\quad k_{ij} - \lambda \delta_{ij} = 0 \qquad i,j = 1, 2, 3, \ldots, m$

This is a determinant equation which can be solved to give the values of λ to be used in the general solution [Eq. (B.4)].

The general solution is then

(B.9)
$$A_i = \sum_{r=1}^{m} B_{ir}Q_r^0 e^{-\lambda_r t}$$

where Q_r^0 can be determined from the initial conditions.

Example

$$\begin{array}{c} \text{Pulse} \\ \text{input} \\ \downarrow \end{array}$$

$$A_1 \underset{k_{21}}{\overset{k_{12}}{\rightleftharpoons}} A_2 \underset{k_{32}}{\overset{k_{23}}{\rightleftharpoons}} A_3$$

The rate equations for this three-compartment model are

$$\frac{dA_1}{dt} + k_{12}A_1 - k_{21}A_2 = 0$$

$$\frac{dA_2}{dt} - k_{12}A_1 + k_{21}A_2 + k_{23}A_2 - k_{32}A_3 = 0$$

$$\frac{dA_3}{dt} - k_{23}A_2 + k_{32}A_3 = 0$$

Equation (B.8) in this case can be written

$$\begin{vmatrix} k_{12} - \lambda & -k_{21} & 0 \\ -k_{12} & k_{21} + k_{23} - \lambda & -k_{32} \\ 0 & -k_{23} & k_{32} - \lambda \end{vmatrix} = 0$$

which has solutions

$$\lambda = 0 \text{ or}$$
$$\lambda^2 - (k_{12} + k_{21} + k_{23} + k_{32})\lambda + (k_{12}k_{32} + k_{12}k_{23} + k_{32}k_{21}) = 0$$

This equation can be solved, and we find

$$\lambda_2 = \frac{1}{2}(p + q)$$

$$\lambda_3 = \frac{1}{2}(p - q)$$

with

$$p = k_{12} + k_{21} + k_{23} + k_{32}$$
$$q = [p^2 - 4(k_{12}k_{23} + k_{21}k_{32} + k_{12}k_{32})]^{1/2}$$

The values of B_{11}, B_{21}, and B_{32} can be found by solving Equation (B.7). In this case Equation (B.7) is

(B.10)
$$(k_{12} - \lambda_r)B_{11} - k_{21}B_{2r} = 0$$

(B.11)
$$-k_{12}B_{1r} + (k_{21} + k_{23} - \lambda_r)B_{2r} - k_{32}B_{3r} = 0$$

(B.12)
$$-k_{23}B_{21} + (k_{32} - \lambda_r)B_{3r} = 0$$

If $B_{1r} = 1$, it is obvious that Equations (B.10) and (B.11) can be solved to give

$$B_{2r} = \frac{k_{12} - \lambda_r}{k_{21}}$$

$$B_{3r} = \frac{k_{23}(k_{12} - \lambda_r)}{k_{21}(k_{32} - \lambda_r)}$$

so our solutions are

(B.13)
$$A_1 = \sum_{r=1}^{3} Q_r^0 e^{-\lambda_r t}$$

(B.14)
$$A_2 = \sum_{r=1}^{2} Q_r^0 \frac{k_{12} - \lambda_r}{k_{21}} e^{-\lambda_r t}$$

(B.15)
$$A_3 = \sum_{r=1}^{3} Q_r^0 \frac{k_{23}(k_{12} - \lambda_r)}{k_{21}(k_{32} - \lambda_r)} e^{-\lambda_r t}$$

To find Q, we need to put in the initial conditions. This example with $k_{32} = 0$ is of course equivalent to the model

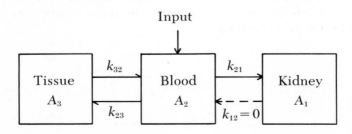

A_1 is designated the kidney, A_2 is the blood pool into which the isotope is injected, and A_3 is the tissue.

If at time $t = 0$, $A_1 = 0$, $A_3 = 0$, and $A_2 = A_2^0$ initially as $k_{12} = 0$ and $\lambda_1 = 0$:

(B.16)
$$0 = Q_1^0 + Q_2^0 + Q_3^0$$

(B.17)
$$A_2^0 = -\frac{Q_2^0 \lambda_2}{k_{21}} - \frac{Q_3^0 \lambda_3}{k_{21}}$$

(B.18)
$$0 = -\frac{Q_2^0 k_{23} \lambda_2}{k_{21}(k_{32} - \lambda_2)} - \frac{Q_3^0 k_{23} \lambda_3}{k_{21}(k_{32} - \lambda_3)}$$

we find that in

$$Q_1^0 = \frac{A_2^0 k_{21} k_{32}}{\lambda_2 \lambda_3}$$

$$Q_2^0 = -\frac{A_2^0 k_{21}(k_{32} - \lambda_2)}{\lambda_2(\lambda_3 - \lambda_2)}$$

$$Q_3^0 = \frac{A_2^0 k_{21}(k_{32} - \lambda_3)}{\lambda_3(\lambda_3 - \lambda_2)}$$

Substituting these values into the general solution,

$$A_1 = A_2^0 \left[\frac{k_{21} k_{32}}{\lambda_2 \lambda_3} - \frac{k_{21}(k_{32} - \lambda_2) e^{-\lambda_2 t}}{\lambda_2(\lambda_3 - \lambda_2)} + \frac{k_{21}(k_{32} - \lambda_3) e^{-\lambda_3 t}}{\lambda_3(\lambda_3 - \lambda_2)} \right]$$

$$A_2 = A_2^0 \left[\frac{k_{32} - \lambda_2}{\lambda_3 - \lambda_2} e^{-\lambda_2 t} - \frac{k_{32} - \lambda_3}{\lambda_3 - \lambda_2} e^{-\lambda_3 t} \right]$$

$$A_3 = A_2^0 \left[\frac{k_{23} e^{-\lambda_2 t}}{\lambda_3 - \lambda_2} - \frac{k_{23} e^{-\lambda_3 t}}{\lambda_3 - \lambda_2} \right]$$

The variations of all three compartments with time can be calculated from these relations.

The correlation between the general solution and the more simple cases can be seen if we put $k_{21} = 0$; in this case we are simply solving for the reversible case:

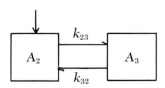

In this case

$$\lambda_2 = k_{23} + k_{32}$$

$$\lambda_3 = 0$$

So

$$A_2 = A_2^0 \left(\frac{k_{23}}{k_{32} + k_{23}} e^{-(k_{23} + k_{32})t} + \frac{k_{32}}{k_{32} + k_{23}} \right)$$

$$A_3 = A_2^0 \left(\frac{k_{23}}{k_{32} + k_{23}} - \frac{k_{23}}{k_{32} + k_{23}} e^{-(k_{23} + k_{32})t} \right)$$

This solution can be obtained much more easily, but these examples show how solutions to all models can be calculated from the general solution.

INDEX

In this index, page numbers in **boldface** indicate illustrations; those followed by the letter "t" indicate tables.